Vital Forest Graphics

Core editorial team

Christian Lambrechts (UNEP)

Mette Løyche Wilkie (FAO)

Ieva Rucevska (UNEP/GRID-Arendal)

Mita Sen (UNFF Secretariat)

Editorial panel

Kevin M. Conrad (Coalition for Rainforest Nations)

Mahendra Joshi (UNFF Secretariat)

Lars Laestadius (Global Forest Watch)

Lars Løvold (Rainforest Foundation Norway)

Claude Martin

Risto Päivinen (The European Forest Institute)

Carsten Smith Olsen (Royal Veterinary and Agricultural University, Denmark)

Language editor: Kieran Cooke
Proof reader: Harry Forster

Cartography

Philippe Rekacewicz assisted by Cécile Marin, Agnès Stienne, Giulio Frigieri, Riccardo Pravettoni, Laura Margueritte and Marion Lecoquierre

Authors and contributors

Frédéric Achard (European Commission Joint Research Centre)

John Bennett (Principal, Bennett & Associates)

Donné Beyer (Octopus Media)

Jim Carle (FAO)

Arnaldo Carneiro (Instituto Socioambiental, Brazil)

Peter Csoka (UNFF Secretariat)

Alberto Del Lungo (FAO)

Frédéric Durand (University of Toulouse, France)

Marianne Fernagut (UNEP/GRID-Arendal)

Lauren E. Haney (UNEP/GRID-Arendal)

John Innes (University of British Columbia)

David Kaimowitz

Marion Karmann (FSC International Centre, Germany)

Ashish Kothari (IUCN)

Ingelin Ladsten (Rainforest Foundation Norway)

Christian Lambrechts (UNEP)

Barbara Lassen (IUCN)

Arvydas Lebedys (FAO)

Mette Løyche Wilkie (FAO)

Claude Martin

Carolyn Marr (Down to Earth, UK)

Lera Miles (UNEP – WCMC)

Mukundi Mutasa (Southern Africa Research and Documentation Centre, Zimbabwe)

Christian Nellemann (UNEP/GRID-Arendal)

Vemund Olsen (Rainforest Foundation Norway)

Martina Otto (UNEP DTIE)

Jari Parviainen (Finnish Forest Research Institute)

Adriana Ramos (Instituto Socioambiental, Brazil)

Philippe Rekacewicz (Le Monde Diplomatique)

Ieva Rucevska (UNEP/GRID-Arendal)

John Sellar (CITES)

Carsten Smith Olsen (Royal Veterinary and Agricultural University, Denmark)

Barbara Tavora-Jainchill (UNFF Secretariat)

Frank Turyatunga (UNEP/GRID-Arendal)

Natalie Unterstell (Instituto Socioambiental, Brazil)

Jo Van Brusselen (The European Forest Institute)

Petteri Vuorinen (FAO)

Monika Weißschnur

Layout and cover: Boris Séméniako

Foreword

The world's forests provide a multitude of environmental, economic and social services, all of which are invaluable in supporting human development. Forests sustain the livelihoods of hundreds of millions of people globally, and contribute directly to the economies of numerous countries.

Yet, about 13 million hectares of forests continue to be lost every year with far reaching consequences in terms of carbon emissions, loss of biodiversity and environmental degradation. Whereas forests and forest soils store more than one trillion tonnes of carbon, current rate of deforestation and forest degradation is responsible for close to 17.4 percent of all anthropogenic greenhouse gas emissions, contributing to climate change. Increasingly, afforestation and reforestation are being promoted as means of climate change mitigation and adaptation. Forests often are at the nexus of the most pressing issues high on the global environmental and sustainable development agenda, namely: climate change, biodiversity loss, poverty eradication, ecosystem management, and environmental governance.

To help communicate the value of forests to policy-makers and the wider public, the United Nations Environment Programme, the Food and Agriculture Organization of the United Nations and the United Nations Forum on Forests Secretariat of the United Nations Department of Economic and Social Affairs joined efforts to analyse, synthesize and illustrate topical forest issues in this new publication, the *Vital Forest Graphics.*

A group of authors from around the world provided case studies and inputs. The publication was edited by a core team, guided by a high-level panel comprising experts from the academia, as well as from leading governmental and non-governmental institutions committed to forest conservation and sustainable management.

This edition of the Vital Graphics series is intended to serve as an advocacy tool to promote conservation and sustainable management of the world's forests through a better and wider understanding of the critical values they provide in support of global ecological stability, economic development and human well-being.

We are pleased to present this publication, and hope that you will find it both informative and thought-provoking.

Achim Steiner	Jan Heino	Jan McAlpine
Under-Secretary-General	Assistant Director-General	Director
Executive Director, UNEP	FAO Forestry Department	UNFF Secretariat

Contents

Forests:
Suppliers of
multiple

Global forest types

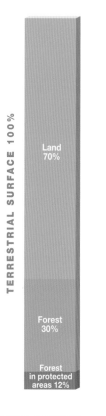

TERRESTRIAL SURFACE 100%

Land 70%

Forest 30%

Forest in protected areas 12%

FOREST OF THE WORLD 100%

Tropical rain-forests 28%

Other tropical forests 19%

Sub-tropical forests 9%

Temperate forests 11%

Boreal forests 33%

Over the last few years, two closely related key environmental issues have been at the top of the environmental agenda: climate change and deforestation. Deforestation, estimated at around 13 million hectares a year, has immediate consequences in terms of increased carbon emissions and loss of biological diversity. Most of the losses in forest cover are taking place in developing countries, in particular in South America, Africa and Southeast Asia. One of the root causes behind deforestation is the weak governance structure for forest conservation and sustainable management of forest resources. This applies particularly to publicly owned forests that represent over 80 per cent of global forest cover.

In order to help address forest governance, it is essential to further the understanding of policy-makers and the general public at large regarding the importance of forests, the underlying causes of their loss and the exciting successful practices available to help conserve them. This calls for a strengthening of the interface between science and policy

← Global forest distribution

services to nature
and humankind

and for efforts to ensure that scientific findings are translated into a common, user-friendly language.

Environmental assessments, such as the Vital Graphics series, are fundamental communication tools that promote interaction between science and various stages of the policy and decision-making cycle. The format of the Vital Graphics series with its extensive graphic component reduces complexity and adds value by summarizing, synthesizing and illustrating critical environmental issues.

Forest issues are wide ranging. In order to effectively raise the awareness and understanding of policy-makers and the general public, Vital Forest Graphics focuses on a number of selected issues that are topical and important.

The publication starts by setting the stage and looking at what defines a forest. It examines changes in forest cover in various parts of the world over the last century. The publication also provides an analysis of the most salient features of the largest forest ecosystems, including the tropical forests of the Amazon, the Congo Basin and Southeast Asia, as well as the boreal forests. The Vital Forest Graphics also analyses the role and importance of forests with regard to the most pressing environmental issues of our time, including: climate change; loss of biodiversity; trade and environment; air pollution; energy and biofuels; agriculture and food security.

In order to further the understanding of the importance of the forests, the publication reviews the main ecological functions they provide in support of human well-being. These include the regulation of ecological processes, including the hydrological cycle and micro-climatic conditions.

Finally the Vital Forest Graphics highlights some of the proven or innovative practices, including legal or economic tools, which have been implemented to help conserve the forests and secure the livelihoods of forest-dependent communities.

Forest definition and extent

*How much forest
is there in the world?
A surprisingly difficult
question to answer*

Defining what constitutes a forest is not easy. Forest types differ widely, determined by factors including latitude, temperature, rainfall patterns, soil composition and human activity. How a forest is defined also depends on who is doing the defining. People living in the British Isles or Scandinavia might identify forests differently from people in Africa or Asia. Similarly, a business person or economist might define and value a forest in a very different way from a forester, farmer or an ornithologist.

A recent study of the various definitions of forests (Lund 2008) found that more than 800 different definitions for forests and wooded areas were in use round the world – with some countries adopting several such definitions at the same time!

It should be kept in mind that different definitions are required for different purposes and at different scales. An assessment focusing on the availability of timber for commercial or industrial purposes may exclude small wooded areas and types of forest not considered to be of commercial value. A definition based on physical characteristics, such as the canopy cover, will most likely be used for an assessment

⬇ The main biomes of the world

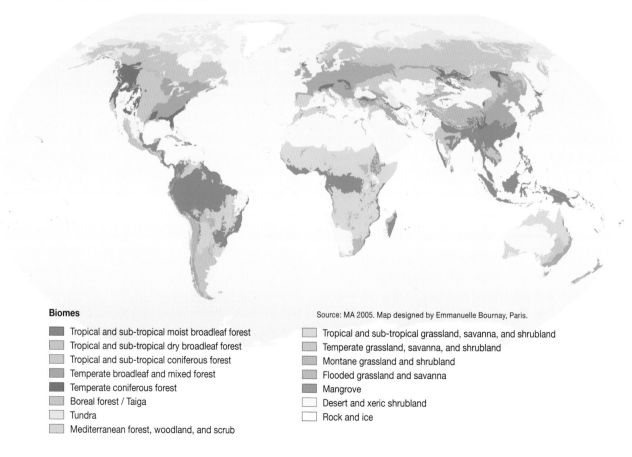

Source: MA 2005. Map designed by Emmanuelle Bournay, Paris.

Biomes

- Tropical and sub-tropical moist broadleaf forest
- Tropical and sub-tropical dry broadleaf forest
- Tropical and sub-tropical coniferous forest
- Temperate broadleaf and mixed forest
- Temperate coniferous forest
- Boreal forest / Taiga
- Tundra
- Mediterranean forest, woodland, and scrub
- Tropical and sub-tropical grassland, savanna, and shrubland
- Temperate grassland, savanna, and shrubland
- Montane grassland and shrubland
- Flooded grassland and savanna
- Mangrove
- Desert and xeric shrubland
- Rock and ice

→ Forest cover varies depending on how it is defined

of the forest extent, whilst a definition based on botanical characteristics, i.e. variety of tree species, will be used for assessing various classes or types of forest. An overall assessment carried out on a regional or global level is unlikely to satisfy more detailed national level requirements. Conversely, a definition developed to suit the needs of any given country is unlikely to be applicable at a global level.

In an attempt to calculate how much forest there is both at regional and global levels some common definitions have been developed. These definitions are generally very broad, in order to encompass all types of forests –from dense, tall forests found in the humid tropics, to temperate and boreal forests and forests in semi-arid and arid regions.

Common Definitions

The Food and Agriculture Organization of the United Nations (FAO) has been assessing the world's forest resources at regular intervals. Its Global Forest Resources Assessments (FRA) are based on data provided by individual countries, using an agreed global definition of forest which includes a minimum threshold for the height of trees (5 m), at least 10 per cent crown cover (canopy density determined by estimating the area of ground shaded by the crown of the trees) and a minimum forest area size (0.5 hectares). Urban parks, orchards and other agricultural tree crops are excluded from this definition – as are agroforestry systems used for agriculture. According to this definition there are at present just under 4 billion hectares of forest in the world, covering in all about 30 per cent of the world's land area (FAO 2006a).

The United Nations Framework Convention on Climate Change (UNFCCC) uses a slightly different ▶

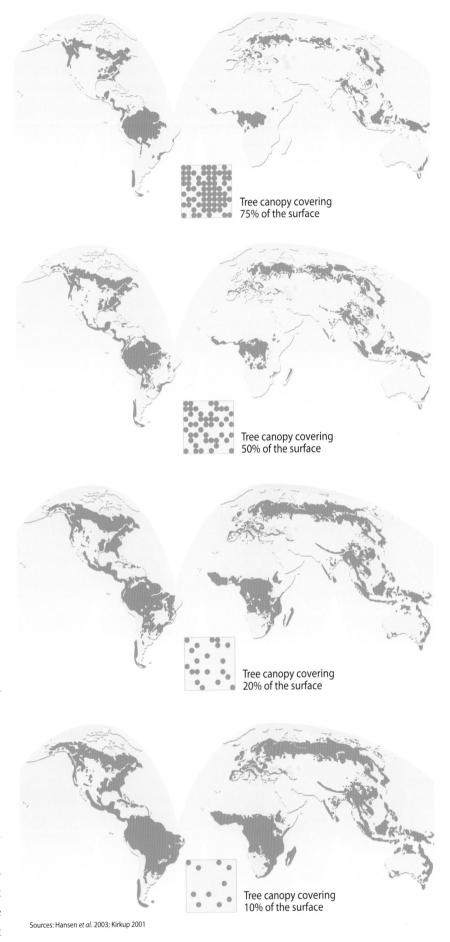

Tree canopy covering 75% of the surface

Tree canopy covering 50% of the surface

Tree canopy covering 20% of the surface

Tree canopy covering 10% of the surface

Sources: Hansen *et al.* 2003; Kirkup 2001

▶ approach. It requests industrialized countries to estimate the forest area according to their own national definitions which should be documented in the greenhouse gas inventory report. For supplementary reporting to the Kyoto Protocol, however, these countries have to apply a forest definition with threshold values within certain parameters; 0.01-1.0 hectares for minimum area, 2-5 meters for minimum tree height and 10-30 per cent for minimum crown cover. The threshold values chosen must be used for all subsequent assessments made during the reporting period and if the definition is different from the definition used by FAO, the country should explain why a different definition was chosen.

The crown cover threshold and the land use criterion are, in most cases, the most critical factors defining forests. The 10 per cent threshold of crown cover encompasses both open and closed forests. The term closed forest refers to areas where tree cover exceeds 40 per cent while the term open forest refers to areas where tree cover is between 10 and 40 per cent. In order to assess the state of the world's closed forests, the United Nations Environment Programme (UNEP) has recently employed other definition criteria,

⬇ Countries with the most forest

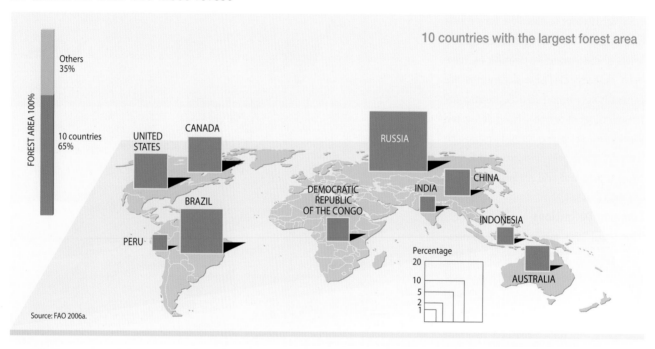

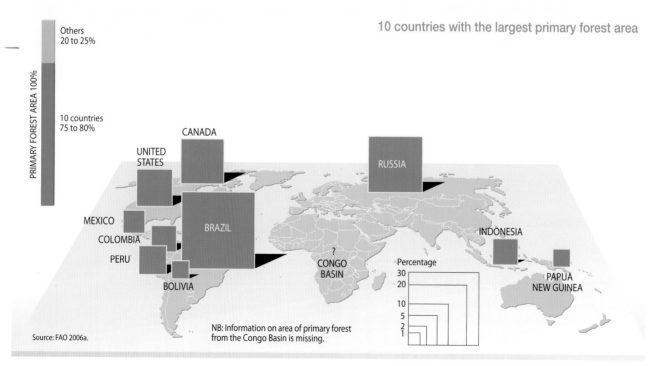

including a minimum crown cover of 40 per cent. It has also used remote sensing to ensure compatibility across countries. According to the UNEP assessment, there were an estimated 2.87 billion hectares of closed forest worldwide in 1995, equivalent to 21.4 per cent of the total land area. Half of this area was located in Russia, Canada and Brazil (UNEP 2001).

Several other regional and global maps and assessments of forests have been produced – often with differing results, reflecting the various definitions and methodologies used and also the differing interpretations made.

Problems which arise in trying to assess the extent of forests worldwide are compounded by the fact that even when using a commonly held definition, data from one country is not necessarily comparable with data from another due to the different methodologies used. For example, the use of satellite imagery might produce very different results to a ground based survey. In addition, remote sensing techniques for assessing forest areas can result in areas used for agricultural purposes or urban development being included rather than excluded in overall calculations of forest area.

In order to help address some of these problems, a new global remote sensing survey of forests carried out by group of agencies led by the FAO is at present being used to assess trends in forest areas over the last 30 years. The survey, which is due to announce its results in 2011, involves all countries and aims to carry out this work in as consistent a way as possible.

A factor not included in the above-mentioned definitions concerns just what a particular forest is made up of. Is it largely composed of indigenous (native) or introduced species? If planted, is it a monoculture – consisting of only one species? The definitions outlined above also exclude the condition of the forest. Is it an undisturbed primary forest, severely degraded forest or something in between? Is the forest healthy or has it been subject to attacks by pests, disease or forest fire, or damaged by wind or air pollution? Area is only one factor in assessing the world's forests: it is also vital to present comparable data on various specific forest types, examine forest health and look at usage and resource values.

➡ See also pages 10, 40-46

⬇ Forest cover in percentage of total land area

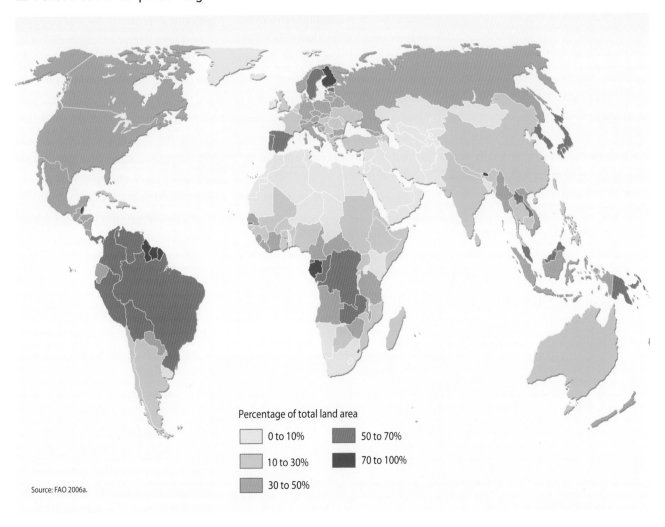

Percentage of total land area

☐ 0 to 10%

☐ 10 to 30%

☐ 30 to 50%

☐ 50 to 70%

☐ 70 to 100%

Source: FAO 2006a.

Forest losses and gains: where do we stand?

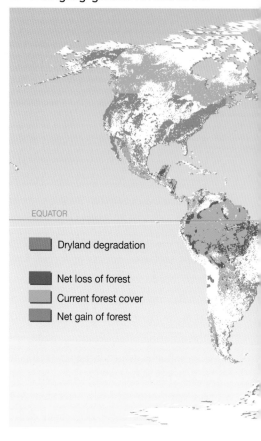

EQUATOR

- Dryland degradation
- Net loss of forest
- Current forest cover
- Net gain of forest

Forests can undergo changes in various ways. Forest areas can be reduced either by deforestation or by natural disasters such as volcanic eruptions or severe mud slides, which can result in the forest being unable to naturally regenerate. Conversely, forest areas can be increased – through afforestation or by the natural expansion of forests

While natural disasters are relatively rare, clearance of forests has been practised throughout documented human history. Prior to the industrial era such clearances were generally part of a relatively slow and steady process (MA 2005) but in recent times the rate of deforestation around the globe has increased dramatically. The Food and Agriculture Organization of the United Nations (FAO) estimates that about 13 million hectares – an area roughly equivalent to the size of Greece – of the world's forests are cut down and converted to other land uses every year (FAO 2006a).

At the same time, planting of trees has resulted in new forests being established while in other areas forests have expanded on to abandoned agricultural land through natural regeneration, thus reducing the net loss of total forest area. In the period 1990-2000 the world is estimated to have suffered a net loss of 8.9 million hectares of forest each year, but in the period 2000-2005 this was reduced to an estimated 7.3 million hectares per year – or an area about the size of Sierra Leone or Panama (FAO 2006). In broader terms, this means that the world lost about 3 per cent of its forests in the period 1990 to 2005; at present we are losing about 200 square kilometres of forest each day.

Unfortunately, very few countries have any estimates of the actual rates of deforestation; even net change estimates are rarely based on regularly conducted assessments – methodologies also differ meaning that estimates have a large degree of uncertainty.

Given the considerable variety in the types of forests and in their characteristics and relative health, the rates of deforestation and net change do not convey the full picture of the changes occurring to forests over time. A net change in forest area may hide the fact that natural forests are being deforested in one part of a country or region while forest plantations are being established in another area. Large scale changes can also happen within the forest area. In some cases natural forests are converted into forest plantations while undisturbed primary forests are being changed into modified or even degraded forests.

For example, forest areas opened by the felling of timber species are likely to be colonised by pioneer tree species, thus changing the forest's composition. It is therefore important not to focus solely on factors such as deforestation rates or net change, but to

⬇ Annual net change in forest area, 1990-2005

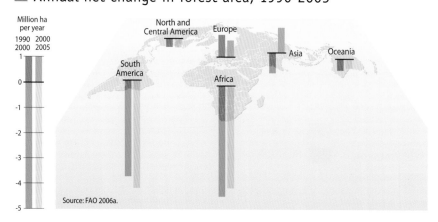

Million ha per year
1990 2000
2000 2005

North and Central America
Europe
South America
Africa
Asia
Oceania

Source: FAO 2006a.

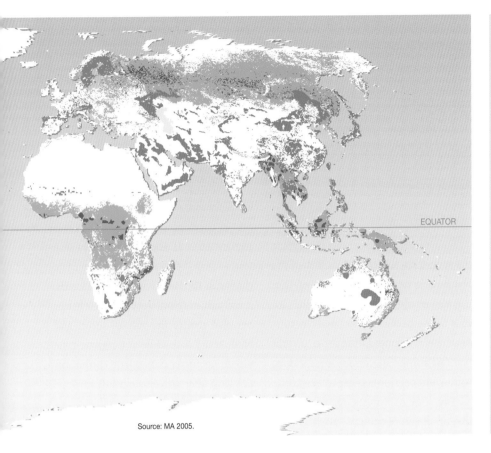

Source: MA 2005.

also look at changes in the characteristics, composition and health of forest ecosystems.

Historically, deforestation has been much more intensive in temperate regions than in tropical regions, with Europe being the continent with the least original forest. However, in the last 50-100 years, the situation has changed; rates of deforestation are now highest in tropical developing countries.

In the period 2000-2005, South America reported the largest net loss of forest, followed by Africa. In the 1990s, Asia had a net forest loss of 800 000 hectares per year. In the period 2000-2005 Asia showed a net gain of forests of around 1 million hectares per year, despite high rates of deforestation in many countries in the region, in particular in Southeast Asia. This net gain is attributed to large-scale afforestation, particularly in China, where there has been an annual increase of more than 4 million hectares. Meanwhile in Europe forest areas continued to expand, although at a relatively slow rate, while North and Central America and Oce-

ania registered a relatively small annual net loss of forests over the 1990-2005 period (FAO 2006a).

The five countries with the largest annual net loss of forest area in the period 2000-2005 were Brazil, Indonesia, Sudan, Myanmar and Zambia. The five countries with the largest annual net gain in forest area over the same period were China, Spain, Vietnam, the United States and Italy. Chile, Costa Rica, India and Vietnam are among the countries which have

recently recorded a change from having a net loss of forests to having a net gain in forest area (FAO 2006a).

Although reasons for deforestation differ from region to region, the most direct cause is generally the conversion of forest areas for agricultural uses, in particular agricultural crops, including annual crops and tree crops, such as orchards and palm oil plantations, as well as for livestock grazing areas. Although harvesting of tropical timber is rarely the main cause of ▶

⬇ Global forest fragmentation

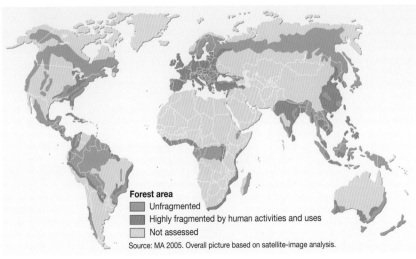

Forest area
- Unfragmented
- Highly fragmented by human activities and uses
- Not assessed

Source: MA 2005. Overall picture based on satellite-image analysis.

▶ deforestation, the establishment of logging roads tends to open up previously closed forest areas and facilitate access which then may lead to the conversion of forest areas to agriculture. Underlying causes of deforestation include population increases leading to increases in demand for land, poverty, lack of enforceable property rights and a lack of incentives to establish proper forest management systems.

Forest degradation often implies a change in the health and vitality of a forest ecosystem but it can also relate to other factors such as changes in the composition of tree species, a loss of biodiversity, a permanent or long term reduction in the crown cover and changes in timber volumes or carbon retention levels. Degradation is often caused by overexploitation of forest areas by humans, including haphazard and badly executed logging operations. It can also be caused by pests and diseases or repeated forest fires. Degradation does not of itself result in the loss of forest area but it is often the first phase of a process which ultimately results in deforestation. No reliable data currently exists on the degree of global forest degradation, due in part to different perceptions of what degradation entails and the lack of adequate assessment methodologies with measurable thresholds and/or the resources needed for their implementation.

At times, the condition of a degraded forest can be improved, either through forest or landscape restoration projects or by natural recovery.

Forest fragmentation can jeopardize the long-term health and vitality of the forest ecosystem. Forest fragmentation can also result in species loss as the size of a forest becomes too small to support a viable population of a certain plant or animal species, or if migratory routes and corridors cease to exist.

The loss of forests results in the loss of all the resources – such as timber, fuelwood and non-wood forest products – and services – such as conservation of soil, water and biological diversity – that a forest provides. Loss of forest also means that the vital role the forest plays in carbon storage and sequestration is no longer possible. Removing forests not only means the loss of this carbon carrying capacity but also frequently means that large amounts of greenhouse gas are suddenly released into the atmosphere through wood burning and clearance activities, compounding climate change problems.

Reducing carbon emissions caused by deforestation and forest degradation in developing countries (REDD) is seen by many as a potentially promising approach in the battle to combat climate change. If the REDD initiative succeeds, it will not only mitigate climate change but also reduce the rate of forest and biodiversity loss while at the same time providing forest-dependent communities with alternative sources of income. On a broader level, it will result in developing countries being paid to conserve and sustainably manage large areas of their forests for the benefit of mankind.

→ See also pages
6, 20, 30, 40, 42, 44

⬇ Changes in area covered by forest, 1990-2005

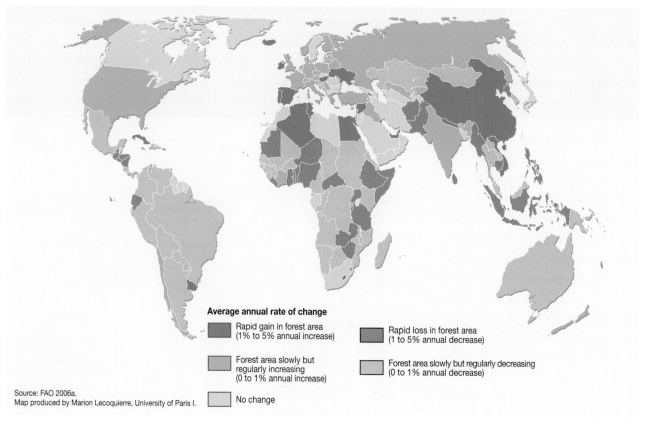

Average annual rate of change

Rapid gain in forest area
(1% to 5% annual increase)

Rapid loss in forest area
(1 to 5% annual decrease)

Forest area slowly but regularly increasing
(0 to 1% annual increase)

Forest area slowly but regularly decreasing
(0 to 1% annual decrease)

No change

Source: FAO 2006a.
Map produced by Marion Lecoquierre, University of Paris I.

⬇ Decreases and increases in forest area in Costa Rica, 1940-2005

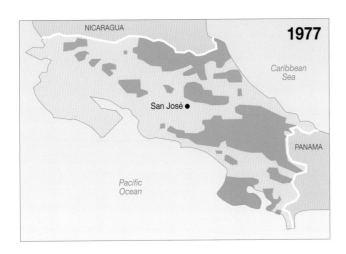

⬇ Conversion of original biomes, 1950-2050

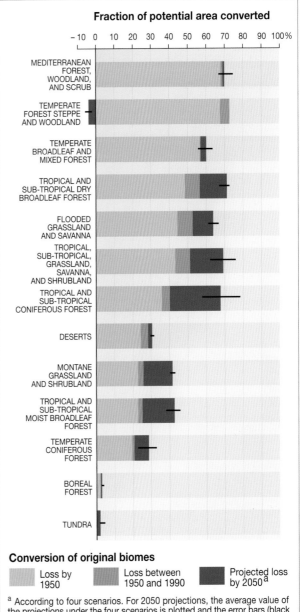

Fraction of potential area converted

Conversion of original biomes

| Loss by 1950 | Loss between 1950 and 1990 | Projected loss by 2050[a] |

[a] According to four scenarios. For 2050 projections, the average value of the projections under the four scenarios is plotted and the error bars (black lines) represent the range of values from the different scenarios.

Source: MA 2005.

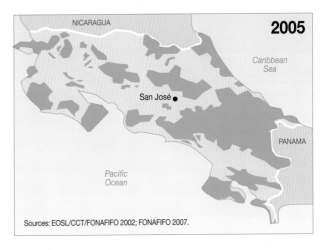

Sources: EOSL/CCT/FONAFIFO 2002; FONAFIFO 2007.

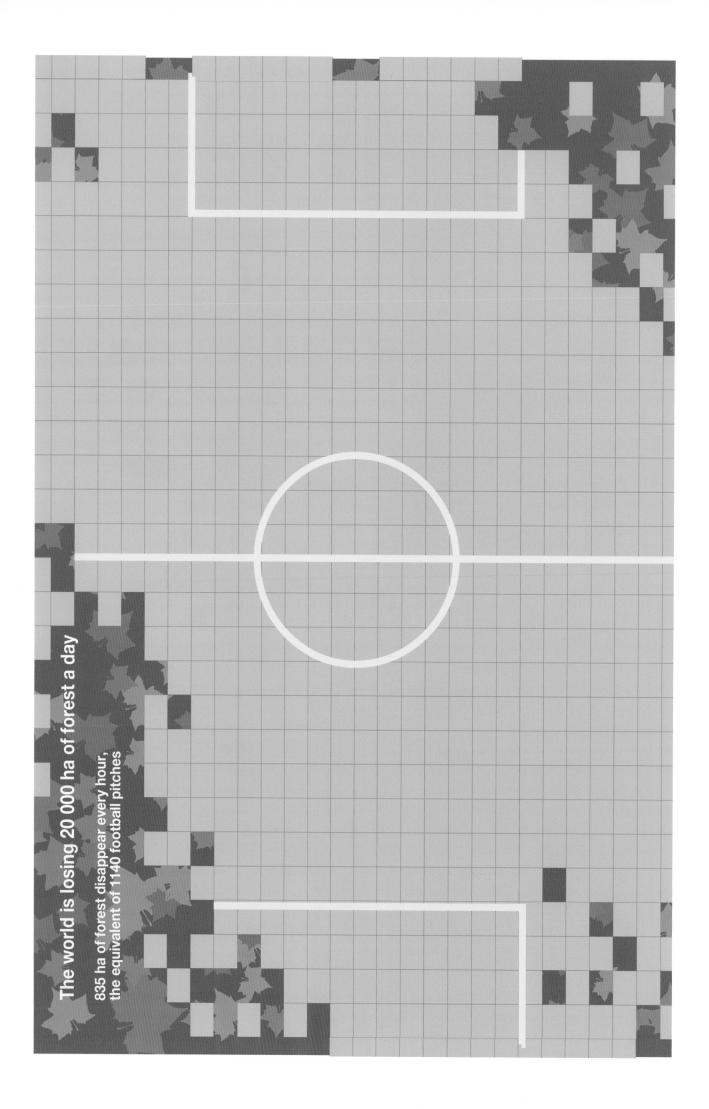

The world is losing 20 000 ha of forest a day

835 ha of forest disappear every hour,
the equivalent of 1140 football pitches

Amazonian deforestation in the global context

Source: Woods Hole Research Center 2007; Amazon Institute for Environmental Research et al. 2006; Reuters 2008.
Research, information collection and elaboration by Giulio Frigieri, University of Bologna, Italy, 2008.

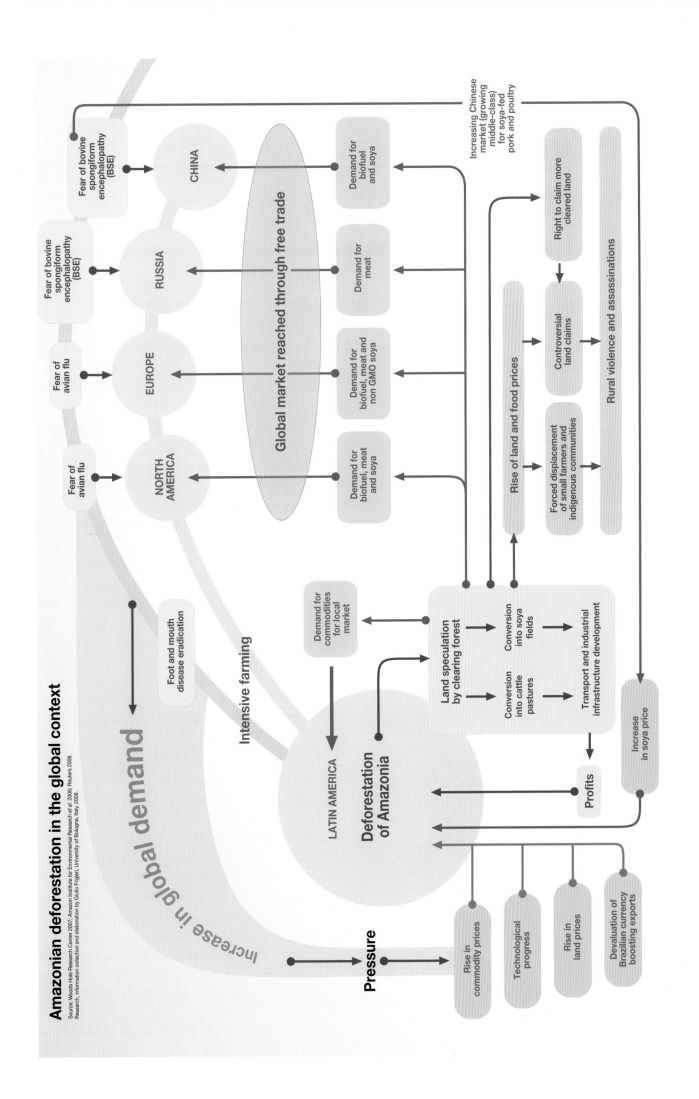

The relationship between indigenous people and forests

More than 1.6 billion people around the world depend to varying degrees on forests for their livelihoods – not just for food but also for fuel, for livestock grazing areas and for medicine. At least 350 million people live inside or close to dense forests, largely dependent on these areas for subsistence and income, while about 60 million indigenous people are almost wholly dependent on forests (World Bank 2006c).

Indigenous forest people use their land in many different ways – for fishing, hunting, shifting agriculture, the gathering of wild forest products and other activities. For them, the forest is the very basis of survival and its resources have to be harvested in a sustainable manner. But when traditional life styles change and, for example, industrial logging or mining takes place, over use of resources can lead to conflict.

Although indigenous people around the world often have very different sets of beliefs and traditions, a special bond with the land is a common factor. For example, the semi-nomadic Matses people of the Peruvian Amazon call the rainforest Titá, or mother (Krogh 2006), referring to Titá as if to a person, who can be happy as well as sad, angry as well as indifferent.

Titá provides the Matses with everything they need – as long as they follow her rules, including never taking more from the forest than is needed and treating all things belonging to it with respect. Traditionally, the Matses perform hunting ceremonies to ask the animal spirits for permission to kill animals for food.

As with the Matses, indigenous peoples' ideas of territory are not only concerned with controlling a geographical area or using forest resources: territory also embodies fundamental aspects of culture and geography (Franky 2000).

Indigenous forest people see themselves as inseparably linked to the forest and everything in it – trees, plants, rivers, animals and mountains. It is impossible, according to community beliefs, to separate any single object or living thing in the forest – such as a particular plant, animal or mineral – from its symbolic position in the cosmology of the people (Olsen 2006). These ideas are expressed through mythology, religious practices, and systems of social regulation, including management of the environment and systems of production and exchange (Sanchez *et al.* 2000).

Because of their special relationship with the land, many indigenous people

⬇ Forest concessions and protected areas, Democratic Republic of the Congo

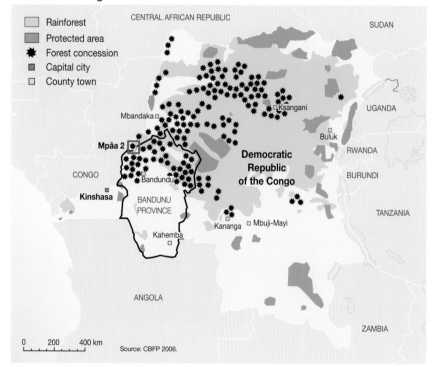

Source: CBFP 2006.

↧ Traditional activities in the village of Mpâa, Democratic Republic of the Congo

Infrastructure
- ≡ Road
- — Footpath

Forest logging
- Forest logging area
- Log storage area

Administration
- Community territory
- Part of territory given as a forest concession
- Village △ Camp

Water
- Main river
- Small river
- Stream
- Water source

Traditional activities
- Cultivated fields
- Harvest area for non-wood forest products
- Fishing area
- × Hunting area
- ∩ Sacred site

Farm

Source: After a map created by Matthieu Yela Bonketo and Barthelemy Boika Mahambi in September 2007; Cercle pour la défense de l'environnement (Ceden), Cenadep; Laboratoire numérique de cartographie participative, Réseau resources naturelles (RRN) and Rainforest Foundation.

0   1   2   3 km

cannot comprehend the idea that forests and land can be bought and sold. However this does not mean they do not have a clear notion of their rights (Odegard *et al.* 2006). The use of certain areas or resources may be granted based on a number of criteria, such as belonging to a particular group, tribe or clan. Land use can also be based on reciprocal agreements with neighbouring groups or individuals.

In many countries, the State is the official owner of most forest areas, even though some of the land may have been inhabited for generations by large numbers of people. In some cases the rights of those people are recog-

nized. In the Philippines for example, land issues in those areas are governed by the Indigenous Peoples' Rights Act. Unfortunately such regulations are often contravened by powerful local interests.

Also, traditional tenure systems are not always recognized by governments, leaving indigenous forest people without formal rights to their territories. This violates the United Nations Declaration on Indigenous Peoples' Rights (UNDIPR) as well as ILO Convention 169 – both of which place a clear obligation on States to legally recognize, demarcate and effectively protect indigenous peoples' territories and natural

resources.

One strategy which is increasingly being used by forest people in order to defend their rights is to provide proof of their residence in, and use of, forest areas. In the Democratic Republic of the Congo, indigenous groups and other forest-dependent communities are participating in the mapping of their traditional territories (FORUM 2007). Such maps are likely to be a vital tool in the future as indigenous people around the world struggle to gain formal recognition of their rights.

→ See also pages 16, 52

Forests sustain livelihoods

Forests play an important role in the livelihoods and welfare of a vast number of people in both developed and developing countries; from urban citizens taking a recreational stroll in a nearby forest to isolated hunter-gatherers who live in and off the forest

The World Bank has estimated that 1.6 billion people around the world depend to some degree on forests for their livelihoods (World Bank 2004). Although only an estimate this clearly indicates that forest dependency is widespread. In developing countries, it is projected that a large number of people will remain at or below poverty levels (Collier 2007). This raises the question of whether forested areas can play a role in poverty alleviation.

A livelihood involves income-generating activities determined by natural, social, human, financial and physical assets and access to these (Ellis 2000). Trees, shrubs, herbs, game and a wide range of other forest products all constitute important natural assets that are harvested in significant quantities by a large number of households across

virtually all forest types (e.g. Scoones *et al.* 1992; Neumann 2000; Cunningham 2001). Such assets therefore make up an important contribution to livelihoods.

Examples are numerous. Fuelwood is an important source of household energy for heating and cooking in many countries. Non-wood forest products, such as bush meat, are important to help meet dietary deficits and a vital source of protein. Medicinal plants from the forest, used either in self-medication or in traditional medicine systems are in many regions the sole or main source of medicinal remedies for maintaining or improving health. Small-scale forest product processing, such as wood carvings or cane furniture, may be an important source of non-farm employment.

Even though forests are often very important to households, there is surprisingly little knowledge on the actual level of household forest income and the role of such income in maintaining livelihoods. Households typically use forest products for subsistence purposes or products are traded in informal markets. Much forest use is therefore not recorded in regular income surveys. However, available evidence indicates that income derived from the forest may constitute 20 per cent or more of total household income,

with the poor the most forest dependent (e.g. Cavendish 2000; Angelsen and Wunder 2003; Vedeld *et al.* 2004).

There is evidence that forests are often of particular importance to women, children and ethnic minorities. For instance, forest foods are crucial to many children (McGregor 1995) and involvement of women in non-timber forest product collection and trade improves intra-household equity (Kusters *et al.* 2006). There are also studies indicating that richer households may be highly forest dependent – though such dependence relates to other sets of products than the ones extracted by poor households. For instance, fuel wood and the use of dung has been found to decrease as income rises in India while fodder and the use of wood for construction increases (Narain *et al.* 2008).

The evidence regarding the role of forests in allowing households to move out of poverty is scant and mixed; there are examples such as the above study from India indicating that income from forests allows households to accumulate assets and escape poverty. However, by way of contrast, figures from Madagascar show that areas there with high forest cover have low densities of people but high poverty rates.

The World Bank and the Food and Agriculture Organization of the United Nations (FAO) have urged that forests can and must play a far bigger role in meeting the United Nations' Millennium Development Goals, including the target of halving extreme poverty by 2015 (World Bank 2004; FAO 2005).

⬇ Level of dependence on forests

Low dependence

Full dependence

Northern consumers in urban areas

Employees in forest industries

Small-scale farmers living outside forests

Urban poor in developing countries

Isolated hunter-gatherer communities

⬇ Forest cover in relation to poverty, Madagascar

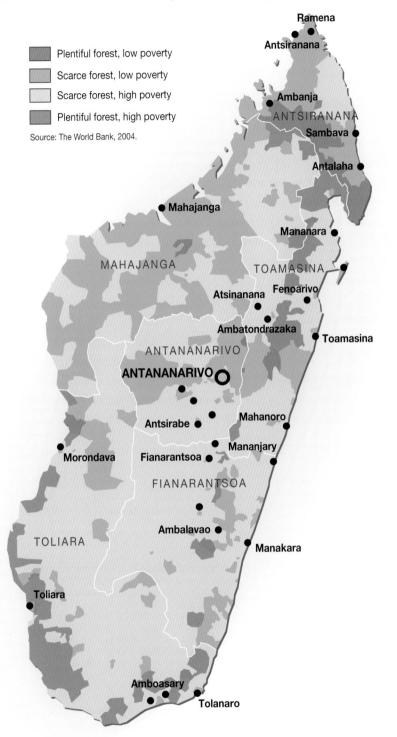

Plentiful forest, low poverty

Scarce forest, low poverty

Scarce forest, high poverty

Plentiful forest, high poverty

Source: The World Bank, 2004.

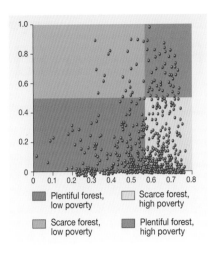

Plentiful forest, low poverty

Scarce forest, low poverty

Scarce forest, high poverty

Plentiful forest, high poverty

ential treatment to influential companies and organizations and promotes corruption. In Honduras, where local communities cannot gain secure rights to the forests in which they live while common, systematic and high-level corruption has characterized the timber industry (Larson and Ribot 2007). There is thus scope for increasing the contribution of forests to poverty prevention and reduction.

Building upon the emerging evidence of the absolute and relative importance of forests and forest products to livelihoods, governments and other development bodies should take action to: make policy reforms in negotiation with small-scale forest users in order to create conducive production conditions, including secure ownership and use rights; revise legislation in order to remove bias against household-level producers and support small-scale commercial units, including community-based forestry. Such initiatives would allow households to actively use forests, enabling them to build up their assets and improve their livelihoods.

There are also international initiatives aimed at improving our understanding of the relation between forests and livelihoods and the impact of policies on such relations, including the Poverty and Environment Network, the Programme on Forests and the International Forestry Resources and Institutions research programme.

Being able to not only harvest forest products but also to transport and sell such products is important for hundreds of thousands of households in order to fully realise the benefits of the forest. However, legislation often discriminates against small forest users, typically by heavily regulating their access rights. It also often gives prefer-

➡ See also pages 14, 32, 52

A hiding place for fighting forces and a refuge for victims

Around the world, conflicts and wars are taking a toll on forests and on the communities that rely on them for their livelihood. Dense forests can serve as hideouts for insurgent groups or can be as a vital source of revenue for warring parties to sustain conflict.

Around the world, conflicts and wars are, directly and indirectly, taking a toll on forests and the communities that rely on forest resources for their livelihood. Dense forests in remote areas can serve as hideouts for insurgent groups. They also provide safe haven for refugees fleeing from conflict. Both cases can result in over-exploitation of forest resources.

Known cases of forests as sites of rebel camps include Colombia where left-wing guerrillas have camps deep in the Amazonian forest and in mountainous forest areas, and the Demo-cratic Republic of the Congo (DRC) where the Garamba forest has been a rebel stronghold for nearly two decades. Many other wars and conflicts, for example, in Cote d'Ivoire, Guinea, Nicaragua, Sierra Leone, and the liberation struggles in southern Africa, were also based in and launched from forest areas.

Forest resources can be a vital source of revenue, used by warring parties to finance and sustain conflict; common conflict commodities are timber and mineral resources extracted from forest areas such as coltan, gold, diamonds

⬇ Forests in narcotics and arms trafficking areas

Altitude Metres
50
100
200
300
500
1.000
2.000
3.000
4.000

Mataven forest
Frequent fires
Coca cultivation
Arial spraying for coca culture destruction
Pollution diffusion
Overfishing
Transboundary traffic (arms, small weapons and narcotics)

××× Closure's project for the Mataven forest
National park
Forest reserve
Western limit of the IUCN priority area for birds, fish, amphibians and floristic protection
Vulnerable ecosystems

Source: UNODC, 2005; data and information collected during various field-trips by the Institute for Environmental Security, The Hague. Map compiled in collaboration with Diana Duarte Rizzolio, UNEP-GRID/Europe.

⬇ Forests affected as hideouts and refuges

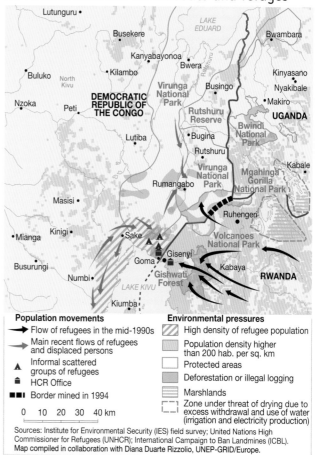

Population movements
➡ Flow of refugees in the mid-1990s
➡ Main recent flows of refugees and displaced persons
▲ Informal scattered groups of refugees
⌂ HCR Office
■■■ Border mined in 1994

0 10 20 30 40 km

Environmental pressures
High density of refugee population
Population density higher than 200 hab. per sq. km
Protected areas
Deforestation or illegal logging
Marshlands
Zone under threat of drying due to excess withdrawal and use of water (irrigation and electricity production)

Sources: Institute for Environmental Security (IES) field survey; United Nations High Commissioner for Refugees (UNHCR); International Campaign to Ban Landmines (ICBL). Map compiled in collaboration with Diana Duarte Rizzolio, UNEP-GRID/Europe.

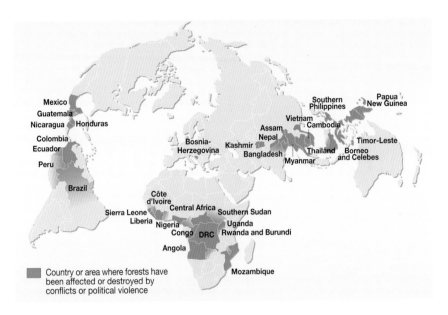

Mexico
Guatemala
Nicaragua Honduras
Colombia
Ecuador
Peru
Brazil
Bosnia-
Herzegovina Kashmir
Bangladesh
Assam
Nepal
Vietnam Cambodia
Southern
Philippines
Papua
New Guinea
Thailand
Myanmar
Borneo
and Celebes
Timor-Leste
Côte
d'Ivoire Central Africa Southern Sudan
Sierra Leone Uganda
Liberia Nigeria Rwanda and Burundi
Congo DRC
Angola
Mozambique

■ Country or area where forests have been affected or destroyed by conflicts or political violence

and other gemstones. However given that logs are bulky and difficult to transport, illegal timber trading in conflict zones can be difficult, requiring access to reasonably good and secure transport systems, as well as access to export markets and consumers.

"Natural resources, including forests, will continue to fuel deadly conflicts as long as consumer societies import materials with little regard for their origin or the conditions under which they were produced. This also makes consumers part of the equation, either knowingly or otherwise" says Michael Renner, a security specialist and senior researcher at the Worldwatch Institute in Washington.

United Nations action to prevent the trade in conflict timber has centred on resolutions, commodity sanctions, and so called smart sanctions, including travel bans and asset freezing of companies and individuals engaged in illegal trading. For more than a decade forest resources were used to finance war in Liberia (1989 to 1996 and 1999 to 2003). In 2003, the UN Security Council called for an import ban on timber products and rough diamonds from Liberia and an end to arms sales to Liberia (UN Security Council 2003). In 2006, following the implementation of new forestry legislation and reforms

by the Liberian government, the timber sanctions were lifted. Subsequently, in 2007, the diamond sanctions were also lifted; however regulations concerning arms sales, travel bans and asset freezes are still in force (UN Security Council 2006).

Important forests

Protected forest areas are seen as a place of refuge by people fleeing from civil unrest, as they tend to be located in remote areas often with dense vegetation where people can hide (Oglethorpe *et al.* 2007). Mwiza Vareriya, caught up in the genocide in Rwanda, described the feelings of those hiding in the Gishwati forest: *"We were very scared, but then we got to the forest, and we felt safer"* (Hanes 2006).

Large-scale cross-border migrations by refugees and internally displaced persons (IDPs) who then settle in forested areas often result in biodiversity loss and deforestation as trees are cut for firewood or shelter and land is cleared for planting crops. In 1994, nearly 1 million refugees from the conflict in Rwanda poured over the border with the DRC and into camps in and around the Virunga National Park (Debroux *et al.* 2007). It is estimated that 50 000 hectares of the park's lowland forest were

affected by woodcutting and animal poaching associated with the humanitarian crisis. An estimated 600 tonnes of wood per day was cut by refugees for fuel, charcoal and shelter, according to the UNEP-World Conservation Monitoring Centre (UNEP-WCMC). International assistance to the refugees included vital food, water and health care. Yet in most cases fuelwood and poles to build shelters were not provided, leaving the refugees with no choice but to exploit local resources.

Sustainable management of forest resources can play a critical role in post-conflict reconstruction and peace building activities. In 2003, the United Nations Environment Programme (UNEP), in its post-conflict environmental assessment of Afghanistan, found that more than 50 per cent of the country's natural pistachio woodlands had virtually disappeared as a result of warfare, civil disorder, institutional collapse and drought. Trees were cut for the illegal sale of timber or to create stockpiles of fuelwood.

In other parts of Afghanistan, the presence of landmines drove farmers to clear forests to grow crops (UNEP 2003). Programmes to restore these devastated areas and create sustainable livelihoods are under way; up to now, the Afghanistan Conservation Corps (ACC) has planted more than 5 million trees and generated income equivalent to over 700 000 labour days (UNEP 2003).

→ See also pages 16, 40

Forests under threat as agricultural

Growing global demand for land for the production of agricultural commodities has resulted in sometimes irreversible changes to the world's forest cover

⬇ Major producers of soya beans and sugar cane

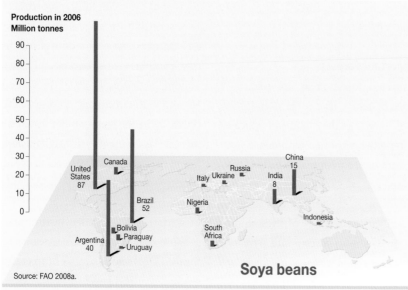

Production in 2006
Million tonnes

Source: FAO 2008a.

Soya beans

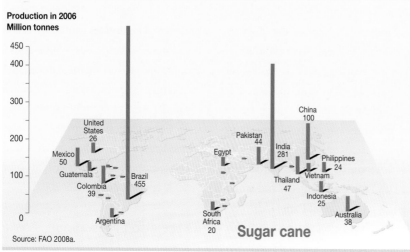

Production in 2006
Million tonnes

Source: FAO 2008a.

Sugar cane

Deforestation is driven by the need for land for uses such as agriculture, agroforestry, human settlements, infrastructure and mining. Some of the most serious deforestation occurs when there are various commodity booms at the domestic and international levels. At such times farmers and large agribusiness enterprises clear forest areas to plant more profitable market crops such as palm oil, rice, sugar cane, banana and soya beans – or forest is cleared for animal grazing. In addition, new road networks mean previously remote areas can be reached and infrastructure built up, often leading to the eventual settlement of former forest lands.

A combination of increase in demands for food, land fertility, rising market prices for commodities and a lack of clear and enforceable ownership rights to forest land results in agriculture being the major cause of deforestation. Such deforestation is often executed through slash and burn practices, with forests and woodlands cut and burned and the land cleared for crop production or for livestock pastures. This method, practised by small-scale farmers for centuries, releases a pulse of nutrients that serve as fertilizers for the soil. However, while traditional slash-and-burn or so called shifting cultivation practices involved the growing of crops for a few years, followed by a fallow period during which the forest

commodities take over

could grow back, more recent forest clearances are often carried out on a much larger scale and aim to establish permanent agriculture.

Forest areas with the most favourable environmental conditions – those with good drainage and soil fertility – are most likely to be converted for agricultural purposes (Kanninen *et al.* 2007; Stickler *et al.* 2007). Tropical forest nations vary greatly in suitability for mechanized agriculture. In some countries such as Malaysia, French Guiana and Cameroon, virtually all forest land has high agricultural potential or high population densities, while in other nations such as Bolivia, Congo, Venezuela and Guyana one third to one half of the forests are unsuitable for agriculture or have low concentrations of farmers (Stickler *et al.* 2007).

To address pressures on forest land the Brazilian government enacted a law in 2006 stipulating that landowners can clear-cut no more than 20 percent of their forests, while the remaining 80 percent has to remain under forest cover. Though this can be seen as an important step in forest conservation, problems over land ownership and lack of clear land titles mean that the law is often not properly implemented.

Increasing populations also raises demand for food and other commodities, requiring ever more land to be used for production (Kanninen *et al.* 2007). At the present time, production of commodities such as soya beans and palm oil are reaching record levels, with world soya bean production in 2006 reaching about 222 million tonnes. Brazil is the world's second biggest soya bean producer, accounting for 23 per cent of the global total. About 220 thousand square kilometres land in Brazil is now planted with soya (FAO 2008a), mostly in the south-eastern part of the country.

Indonesia and Malaysia are major producers of palm oil: in 2006 these two countries accounted for 85 ▶

⬇ Major producers of palm oil and beef

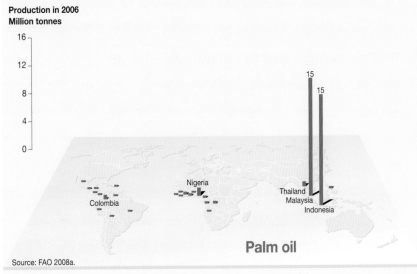

Production in 2006
Million tonnes

Source: FAO 2008a.

Palm oil

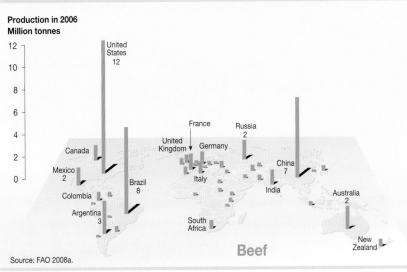

Production in 2006
Million tonnes

Source: FAO 2008a.

Beef

per cent of total world production and 88 per cent of global exports (FAO 2008). Over the past decade, the area covered by oil palms in Indonesia has quadrupled, covering 4.1 million hectares in 2006 (FAO 2008).

Though palm oil tends to be associated mainly with the food and cosmetic industry, its applications are far wider, taking in, for example, the textile industry and more recently – with increased production of bio-fuels – the transportation sector. Many important organizations associated with palm oil, such as the Malaysian Palm Oil Council, regard palm oil as a second-generation bio-fuel which could, through upgraded technologies, become an important source of energy and support the future expansion of the bio-

⬇ Drivers of forest conversion

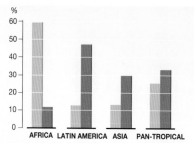

Changes between 1990 and 2000 (in percentage)

▨ Direct conversion of forest area to small-scale permanent agriculture

▨ Direct conversion of forest area to large-scale permanent agriculture

Source: FAO 2001a.

fuel business. Continuing high world prices for palm oil are leading to rapid expansion in areas planted with the crop (Kanninen et al. 2007) and the conversion of forest to oil-palm plantations is a significant contributor to deforestation in some countries.

In Latin America, cattle ranches are expanding rapidly (FAO 2007a) and, according to one study, accounted for an estimated 70 per cent of deforestation in Brazil in 2007 (Malhi et al. 2008). While the average size of a cattle ranch in Brazil is 24 000 hectares, some are as large as 560 000 hectares: in the Brazilian Amazon region, ranches cover an area of at least 8.4 million hectares in total.

The Amazon is now part of a national and international economy which, through globalization, is responding to market demands, accelerating the rate at which agricultural crops and cattle ranching are replacing or impoverishing native forests (Nepstad et al. 2006). The expansion of these agricultural industries in Bra-

⬇ Deforestation benefits...

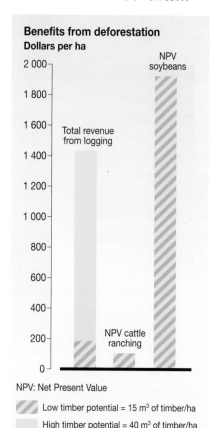

NPV: Net Present Value

▨ Low timber potential = 15 m³ of timber/ha

☐ High timber potential = 40 m³ of timber/ha

Sources: Moutinho and Schwartzman, 2005.

⬇ What is becoming of the Amazonian forest?

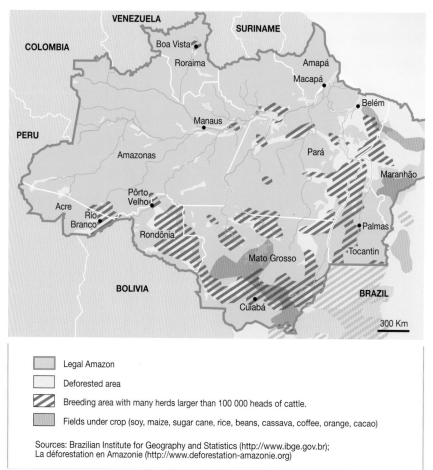

☐ Legal Amazon

☐ Deforested area

▨ Breeding area with many herds larger than 100 000 heads of cattle.

▨ Fields under crop (soy, maize, sugar cane, rice, beans, cassava, coffee, orange, cacao)

Sources: Brazilian Institute for Geography and Statistics (http://www.ibge.gov.br);
La déforestation en Amazonie (http://www.deforestation-amazonie.org)

zil is caused by a complex set of factors including low land prices, devaluation of the Brazilian currency (the Real), and improved transportation infrastructure and production systems. Brazil has also been quick to respond to new export opportunities capitalizing, for example, on meat exports at times when outbreaks of diseases like foot-and-mouth are present in other regions and markets (Nepstad *et al.* 2006).

Many of these trends in land use and agriculture tend to support the neoclassical economic theory that market forces will allocate land to the economically most efficient use (Dadzie 2006-2007). Yet this idea is being challenged with the debate on deforestation at present focussing on the loss of vital ecological services, the economic values of which are still to be fully estimated.

An example of this is the Mabira forest reserve on the shores of Lake Victoria in Uganda. The forest is home to valuable wildlife, serves as a timber resource and provides a number of ecosystem services – yet one third of the reserve has been allocated to sugarcane production.

In the short-term, the growing of sugarcane appears to generate more economic benefits than maintaining the forest reserve intact. Yet over the lifetime of the timber stocks – about 60 years – the benefits that could be derived from the forest would exceed those of planting sugarcane (Moyini *et al.* in press). It should be noted however that such cost benefit analyses differ from area to area and region to region and a thorough evaluation of resources and services is vital before decisions on clearing forest lands are taken.

➡ See also pages 10, 42, 44

⬇ Growth of cattle breeding in Amazonia

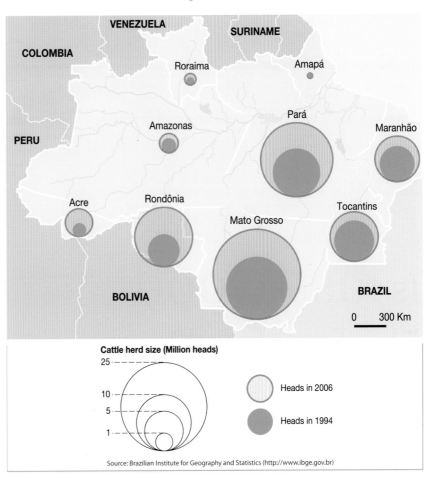

Source: Brazilian Institute for Geography and Statistics (http://www.ibge.gov.br)

⬇ Trends in production and exports of soya beans

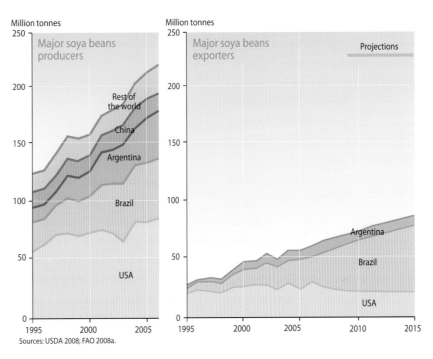

Sources: USDA 2008; FAO 2008a.

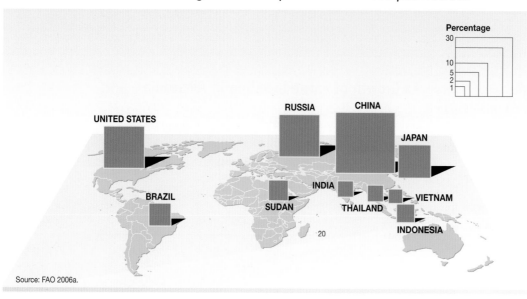

Source: FAO 2006a.

Is fast-wood like fast-food?

The main rationale for forest plantations is that they not only cater for growing demand for forest products but also contribute a vital source of revenue in poor regions. Forest plantations can also help to relieve pressures on natural forests and provide a wide range of social, economic and environmental benefits. Yet in some cases they have negative impacts on biodiversity. In those instances ecosystems are destroyed ultimately leading to the loss of revenue sources and aggravating social problems.

In 2005, 2.8 per cent of total global forest cover was made up of productive forest plantations, amounting to an area of approximately 110 million hectares (FAO 2006a). Productive forest plantations are a source of various goods and services: they provide timber, fuelwood, charcoal, raw materials for panel production, pulpwood for cellulose and paper production, as

Forest plantations meet an increasing proportion of the growing demand for wood products. Yet in some countries around the world, large monocultural fast growing tree plantations have replaced indigenous vegetation cover

well as a wide range of bulk non-wood products such as gum arabic, rubber and cork.

According to the FAO, there was an increase of approximately 40 per cent in the area of the world's forests plantations with productive functions in the 15 years from 1990 (FAO 2006a). Though the majority of these forests is devoted to production of sawn logs and veneer, recently there has been a shift to other products, with about 40 per cent

growth in the areas of planted forests catering for the pulpwood and fibre industries (FAO 2006b). Studies indicate that this growth – a marked trend in recent years – is set to continue (e.g., MA 2005; Down to Earth 2006).

While there has been little growth in Europe in pulp wood plantations and associated paper related activities, the area in Asia doubled between 1990 and 2005 due to large scale planting in China, Indonesia and Vietnam. (FAO 2006b). The trend has been for paper production to shift to southern tropical countries, where the climate is more suitable for fast growing trees. Moreover, land in these regions is more affordable, wages are low and there are generally few barriers in the way of the industrial expansion (Urgewald 2007).

The concern now is that the development of giant pulp mills, each capable of producing as much as 1.2 million tonnes of pulp per year, are going to require yet

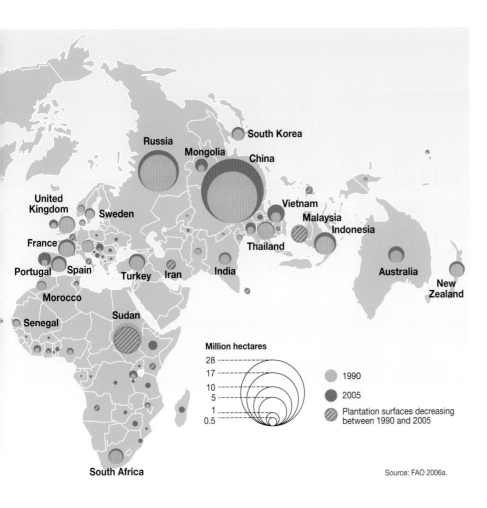

Million hectares
28
17
10
5
1
0.5

1990
2005
Plantation surfaces decreasing
between 1990 and 2005

Source: FAO 2006a.

perate species generally need 60 to 150 years to mature.

In Brazil, China, Indonesia, Thailand, Uruguay and many other countries, concerns have been raised about the impact of fast-wood plantations on biodiversity; such plantations often result in ecosystems being fragmented while hydrological cycles are in some cases disrupted and soil fertility levels reduced (Urgewald 2007).

While there are standards and guidelines for responsible management of forest plantations, progress in implementing them is still slow. The Forest Stewardship Council (FSC), for example, has been able to certify only about seven million hectares of the world's tropical forest plantations. The figure is even smaller for forest plantations established for pulp production (FSC 2008).

→ See also pages 54, 60

more planted forests, particularly in China, the world's biggest paper producer (Urgewald 2007). In order to satisfy the needs of its national pulp industry, the Indonesian government has set an initial target of developing 5 million hectares of industrial wood plantations by 2009 (Down to Earth 2006). In 2005, there were 3.4 million hectares of industrial forest plantations in Indonesia (FAO 2006a).

The way in which paper production has led to the growth of forest plantations has caused concerns. Though most governments subscribe to the idea that natural forests should not be replaced by forest plantations, conversion to forest plantations accounts for six to seven per cent of natural forest losses in tropical countries (Cossalter and Pye-Smith 2005). Indonesia is one of the countries where there has been a massive conversion of tropical forest to industrial tree plantations (Cossalter and Pye-Smith 2005).

Fast growing industrial hardwoods – often referred to as fast-wood – such as eucalyptus and acacia can be harvested in less than 10 years while other plantation species, such as tropical pine and teak, are managed for more than 20 years before harvesting while tem-

↓ Changes (in %) in area of productive forest plantations

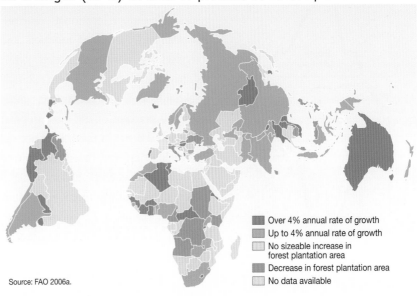

Source: FAO 2006a.

■ Over 4% annual rate of growth
■ Up to 4% annual rate of growth
□ No sizeable increase in forest plantation area
■ Decrease in forest plantation area
□ No data available

Changing trends
in forest products trade

*The international trade
in forest products
has undergone considerable
changes, due to emerging
markets, new investment
strategies and diversification
in products*

⬇ Trends in net trade
of forest products

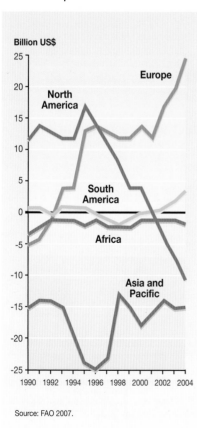

Source: FAO 2007.

In 2004, trade in wood-based forest products accounted for an estimated 3.7 per cent of the world trade in commodities, valued at US$327 billion (FAO 2007). Global trade in forest products takes place mostly within and among countries in Europe and North America and Asia and the Pacific (FAO 2007). Europe accounts for nearly half of the world's trade in forest products with imports of US$158 billion and exports of US$184 billion (FAO 2007).

Global trade in non-wood forest products (NWFPs) such as bamboo, mushrooms, game, fruit, medicinal plants, fibre, gums and resins, has recently been estimated at approximately US$11 billion per annum (FAO 2007).

In addition, the economic value of social, leisure, educational, environmental and medicinal products and services derived from forests, while difficult to estimate, is increasingly acknowledged. For example, more than 50 per cent of the most popular prescribed drugs in the US are derived from natural forest based compounds (Groombridge and Jenkins 2002). Most significantly in recent years international attention has been increasingly focused on the crucial economic and environmental role forests play as a carbon reservoir and sink for mitigating climate change.

Global wood reserves are estimated at 400 billion cubic metres, with gross annual output amounting to roughly 3.5 billion cubic metres of roundwood in 2006 – or about one per cent of global reserves. Half of produced roundwood is used for industrial purposes or in construction and processing (industrial roundwood), the rest as an energy source (woodfuel), mainly in developing countries. In 2004, global output of industrial roundwood was about 1.6 billion cubic metres, of which only 7 per cent was exported (FAO 2007). The bulk of the production was consumed locally or processed into secondary products.

Over the last ten years global demand for roundwood has been growing by about one per cent a year. However the roundwood market is changing substantially, driven by strong demand in emerging economies.

In Europe in 2006 there was a large increase in the production and consumption of sawnwood, resulting in a substantial price rise. Meanwhile in North America the sawnwood market plummeted, due to an economic and housing market slowdown.

Shift in exports and imports

The nature of forest product exports has changed in recent years, with exports of primary products – logs

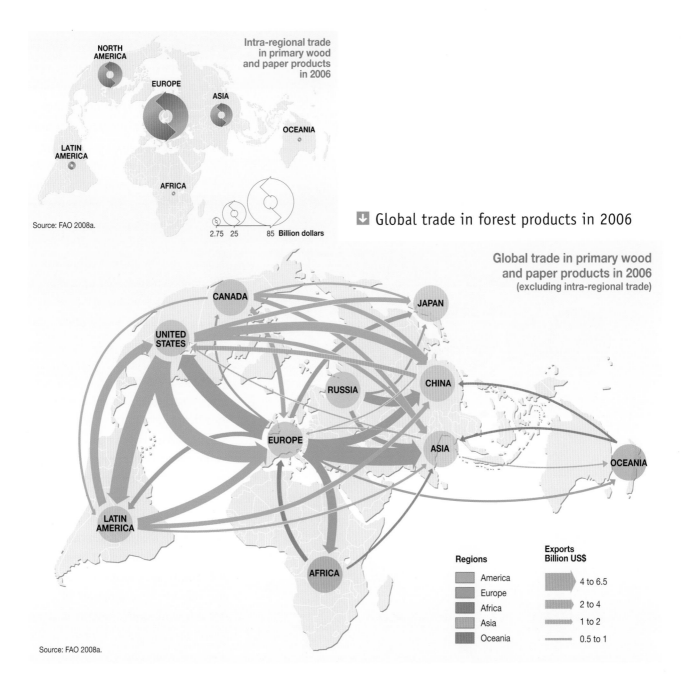

Intra-regional trade
in primary wood
and paper products
in 2006

Source: FAO 2008a.

2.75 25   85 Billion dollars

⬇ Global trade in forest products in 2006

Global trade in primary wood
and paper products in 2006
(excluding intra-regional trade)

Source: FAO 2008a.

Regions

	America
	Europe
	Africa
	Asia
	Oceania

**Exports
Billion US$**

	4 to 6.5
	2 to 4
	1 to 2
	0.5 to 1

and sawnwood – being overtaken by secondary processed wood products (SPWPs), such as furniture and pre-fabricated wooden buildings. In Latin America, only one per cent of logs are exported whole. In Asia, the exports in primary products dropped from 7 per cent to 4 per cent in 2007 (UNECE/ FAO 2007). Some central and western African countries, however, still export logs in significant quantities, making up between 20 to 30 per cent of total production, and mainly destined for Europe and China. This is most likely due to bans on log export placed by countries, such as Brazil and Indonesia, where only added valued products,

including sawnwood and SPWPs, may be exported (UNECE/FAO 2007; WWF 2005). The Russian Federation has recently emerged as the largest exporter of industrial roundwood, accounting for 35 per cent of global trade in 2004, amounting to 42 million cubic metres.

The largest importers of forestry products in general remain the developed countries, led by the United States, Japan and the European Union. However China is catching up with the developed nations in terms of the import and consumption of forestry products. In Scandinavian countries, the Russian Federation and Canada, the domestic output is generally suf-

ficient to meet national demand. Nevertheless, these countries continue to import roundwood.

New investments

Direct foreign investment has boosted the development of wood process-ing industries through technology transfer, infrastructure development and improved access to global mar-kets (FAO 2007). Such investments have been driven by a range of factors including low labour and produc-tion costs, advances in education and research, incentives for foreign invest-ment and a growth in domestic econo-mies. Proximity of forest resources ▶

Production, imports and exports of selected forest products

Paper and paper boards

Tonnes
165 000 000
100 000 000
50 000 000
20 000 000
5 000 000

● Production and imports
● Exports

Industrial roundwood

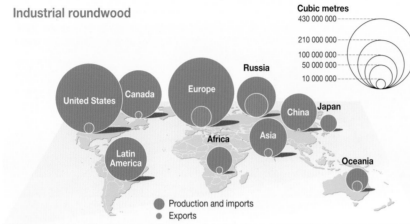

Cubic metres
430 000 000
210 000 000
100 000 000
50 000 000
10 000 000

● Production and imports
● Exports

Wood based panels

Cubic metres
100 000 000
50 000 000
20 000 000
5 000 000

Source: FAO 2008a.

● Production and imports
● Exports

– apart from fuelwood – coupled with a more competitive operating climate, has prompted a shift in investment strategy. Rather than harvesting natural forests, investors have now turned to large-scale forest plantations (Kersanty 2006). Such enterprises are developing fast, particularly in developing tropical countries, with some governments offering subsidies. The ten countries with the largest forest plantation area are supplying just under half of global demand for industrial roundwood. The development of forest plantations, however, poses problems when they replace natural forests and other valuable ecosystems.

Emerging economies

The spectacular recent economic growth of China and, to a lesser extent, Southeast Asia, has changed the nature of the global forestry market, which is now gradually moving from west to east. Chinese processing plants, located in the south of the country, are among the largest in the world. China imports and processes roundwood from Africa and South and Southeast Asia and also imports large amounts of coniferous roundwood from the Russian Federation. In 2004, China became the world's top importer of industrial roundwood and, at the same time, a major exporter of wood panels, paper and carton board.

Illegal trade

The illegal harvesting and trade in forest products is pervasive and often involves unsustainable forest practices which cause serious damage to forests, to forest dependent people and to the economies of producer countries. Illegal trade also increases the cost of forest management and accentuates market distortions (UNFF 2004c). In 2006, the World Bank estimated the annual global losses from illegal cutting of forests in terms of market value at more than US$10 billion,

▶ – once a major consideration – is no longer an important factor as the cost of transporting logs over long distances is largely offset by the low cost of processing. (FAO 2007). Multinational companies are the main players in the timber trade in terms of volume and value exports.

Over the past decade, wood processing industries have sprung up in China and in the countries of Eastern Europe, making them the new players in the global market for production and export of SPWPs (FAO 2007).

The overall drop in the price of raw forest materials over the last ten years plus annual losses in government revenues of about US$5 billion (World Bank 2006). However, the clandestine nature of such illegal trade makes it very difficult to estimate its true scale and value (UNFF 2004c).

Looking forward...

In today's globalized world, the impacts of growing and changing patterns of consumption and production of forest products, combined with changes in forest management, are rapidly transmitted from one region to another (UNFF 2004c). Environmental standards and international regulations – either in legally-binding form or merely laid down as voluntary agreements – governing the management of forests have recently become more strict and this has had an impact on the international market for forest products. While the overall price of forest raw materials has dropped, the price of certified timber has increased.

Concerns over unsustainable forestry practices have led to the creation of certification schemes – market-based mechanisms which aim to promote environmentally sustainable and socially responsible forestry practices. While market demand for certified timber has led big firms to adapt to certification principles and criteria, this has not been possible for many small land-owners and community-based forestry enterprises. As a consequence, such enterprises have often found they are excluded from international markets for certified timber and their activities become restricted to less profitable domestic sales.

International trade can foster economic development and create incentives for sustainable forest management. Equitable access to well-functioning markets can promote sustainable resource management and contribute to poverty alleviation, rural development and the strengthening of forest-based communities. As the global trade in forest products continues to grow, it is important that policies are created and enforced which guarantee trade is not linked to unsustainable resource depletion and illegal practices. If such a course is followed, then trade and sustainable development can become mutually supportive.

Wood products rise

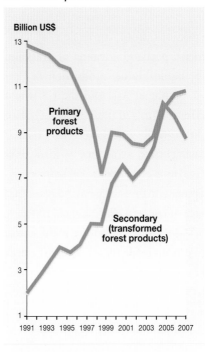

Source: FAO 2007.

Trends in commodity prices

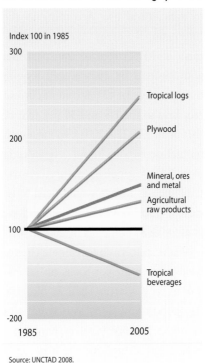

Source: UNCTAD 2008.

Global wood consumption

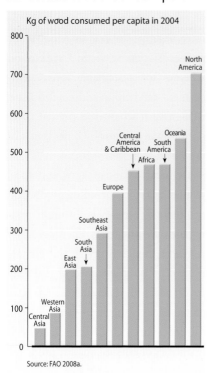

Source: FAO 2008a.

Clearing forests
for biofuels

*To meet growing global energy demand, forest resources
are increasingly exploited and forests are clear-cut
to pave the way for biofuel crops*

Fuelwood and charcoal from forests have long provided energy for heating, cooking and industry. Almost 90 per cent of the wood harvested in Africa, and 40 per cent in Asia and the Pacific, is used for fuel (FAO 2006a). Wood pellets, typically produced in North America and Europe from sawdust and other timber by-products, are increasingly used in stoves, boilers and power stations (Peksa-Blanchard *et al.* 2007). New technologies are also being developed in connection with the production of liquid and gaseous fuels from wood products (FAO 2008).

This growing demand for wood products requires careful management of forest resources in order to minimize negative impacts on biodiversity and ecosystem services. Certification schemes for sustainable forest management can help to address this issue, though such schemes at present cover only a small portion – 7.6 per cent – of the world's forested areas.

Demand for land for agricultural and plantation crops, including production of biomass for energy, is putting increasing pressure on forests. Energy security concerns, high oil prices and climate mitigation policies aimed at replacing fossil fuels with renewable energy, have all led to a greater interest in biofuels. The transport sector is using increasing quantities of ethanol, mainly produced from sugar cane, corn and cassava, as

a substitute for petrol (gasoline), and biodiesel, produced from plantation crops such as oil palm, coconuts and avocados. Fuel ethanol production tripled between 2000 and 2007, mainly in the United States and Brazil, while biodiesel output expanded even more rapidly over the same period, from less than 1 billion to almost 11 billion litres per year (IEA 2004). The liquid biofuel market has been stimulated by growing demand in particular in China and Brazil, and by recent and anticipated legislation in the US and Europe that sets ambitious goals for this sector.

Increasing demand for biofuels has led to more land being converted to agricultural use. According to the International Energy Agency, replacement of 10 per cent of transport fuel with biofuels by 2020 would require the equivalent of 43 per cent of current cropland in the US and 38 per cent of that in the EU (IEA 2004). To meet such biofuel demand without jeopardising global food supplies, natural ecosystems such as forests would need to be cleared (Eickhout *et al.* 2008; Fargione *et al.* 2008; Kanninen *et al.* 2007).

While concerns over climate change have led to a greater focus on renewable energy sources such as biofuels, the carbon emissions associated with deforestation are greater than those avoided by using biofuels from agricultural crops. (Righelato and Sprack-

**Solid biomass consumption in 2003
in % of the total energy consumption**

- ☐ 0 or not significant
- ☐ 2 to 10%
- ■ 10 to 40%
- ■ More than 40%
- ☐ No data

Source: World Resources Institute (WRI) searchable database.

len 2007). It is estimated that carbon emissions from conversion of rainforest, peatland, savanna or grassland for liquid biofuel production in Brazil, Southeast Asia and the US would release 17 to 420 times more CO_2 than the annual greenhouse gas (GHG) savings from avoidance of fossil fuels (Fargione *et al.* 2008). Oil palm development on forested peatland is another area of concern (Parish *et al.*, 2008).

Agriculture is also the main source of nitrous oxide (N_2O), a potent greenhouse gas released by nitrogen fertilizers. In particular, the N_2O emissions associated with rapeseed and maize cultivation for biofuel production outweigh the carbon savings from fossil fuel substitution (Crutzen *et al.* 2008).

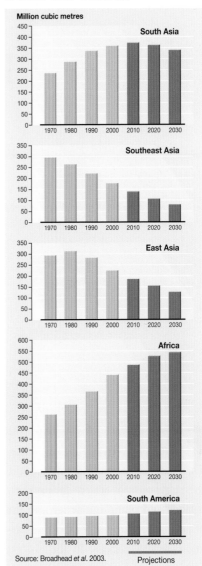

Woodfuel consumption trends and outlook

Million cubic metres

South Asia
1970 1980 1990 2000 2010 2020 2030

Southeast Asia
1970 1980 1990 2000 2010 2020 2030

East Asia
1970 1980 1990 2000 2010 2020 2030

Africa
1970 1980 1990 2000 2010 2020 2030

South America
1970 1980 1990 2000 2010 2020 2030

Source: Broadhead *et al*. 2003.   Projections

Population relying on biomass for cooking and heating

Million

■ 2000
■ 2030

CHINA INDIA OTHER ASIA AFRICA LATIN AMERICA

Source: IEA 2002.

Clearly further analysis on these questions is needed.

The growing debate about the costs and benefits of biofuel production calls for a cautious approach toward further incentives for this sector. It is clear that bioenergy from agricultural crops presents certain opportunities for climate change mitigation, energy security and development. However other options aimed at maximizing the use of agricultural waste products, increasing yields on existing agricultural land and using degraded land for agriculture or forestry are also worth further consideration.

➜ See also pages 20, 24

Global biofuel production

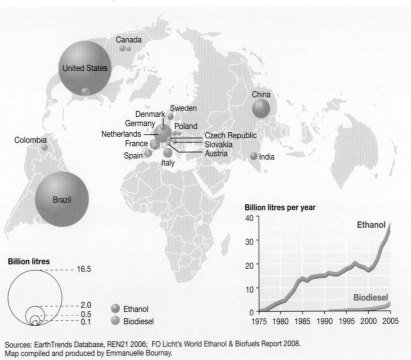

Billion litres
- - - 16.5
- - - 2.0
- - - 0.5
- - - 0.1

● Ethanol
● Biodiesel

Billion litres per year

Ethanol

Biodiesel

1975 1980 1985 1990 1995 2000 2005

Sources: EarthTrends Database, REN21 2006; FO Licht's World Ethanol & Biofuels Report 2008.
Map compiled and produced by Emmanuelle Bournay.

Forests are a key source

Forests provide numerous ecological services. They regulate hydrological cycles, stabilize natural landscapes and protect soils and water courses. The level of these services depends on the location, type and use of forests

Forests interact closely with the water cycle. Trees act like pumps. They help water percolate into the soil, the storehouse of water, during rainy periods and pump out water into the atmosphere, through evapo-transpiration, in dry periods. In the process forests help keep the hydrological cycle "alive".

Deforestation of the entire Amazon basin – a catastrophic scenario – would result, according to projections, in regional decreases of precipitation and evaporation, potentially leading to sustained desertification. Already deforestation is causing fundamental changes in the Amazon climate and hydrology cycle, with possible implications for regional ecosystem dynamics and the global climate (Chagnon and Bras 2005).

Forests can regulate groundwater levels and increase drainage of soils where the water table is close to the surface. If there are salts in the upper soil layers, then removal of forests can result in raised groundwater levels and the movement of salts into the rooting zone of plants (FAO 2008c). In the Murray-Darling Basin in Australia, deforestation and irrigation are the main causes of land salinization affecting some 300 000 hectares (Kabat *et al.* 2004).

Forests can also increase water yields, particularly in tropical cloud forests occurring at relatively high elevations where humidity levels can reach 100 per cent. Large amounts of water are deposited directly onto the vegetation, and excess water is more or less constantly dripping from the leaves to the ground below.

The year-round supply of unpolluted water from cloud forests is a vital resource in many regions. For example, the cloud forests of La Tigra National Park in Honduras provide more than 40 per cent of the water supply for the capital city, Tegucigalpa and its 850 000 people. Other capitals where cloud forests augment water supplies include Quito in Ecuador – a city of 1.3 million people – and Mexico City with its 20 million people (Bubb *et al.* 2004).

Forests also contribute to the maintenance of good water quality. They minimize soil erosion and mitigate flash water flows that cause erosion downstream. In turn this reduces levels of sediment in water bodies such as wetlands, ponds and lakes, streams and rivers. Forests also trap or filter some water pollutants. As water quality levels around the world deteriorate and the cost of filtration facilities remains high, several municipalities have decided to invest resources in the conservation of water catchment areas, including protected forests.

Approximately 9 million people in New York City and nearby areas enjoy access to clean, inexpensive drinking water. About 90 per cent of that water is drawn from the Catskill/Delaware watershed where the abundant forest reserves, as well as soil with adequate carbon levels, provides excellent conditions for natural filtration (WRI 2008a).

⬇ The world is losing its mangroves

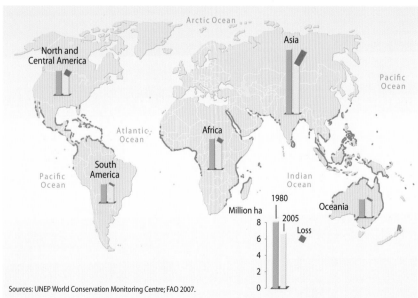

Sources: UNEP World Conservation Monitoring Centre; FAO 2007.

# of ecological services

Flood protection

Forests can also protect against flooding. While forest cover will not appreciably reduce the total amount of water moving into water courses from a large storm event, they can influence the degree of flooding and flood damage (FAO 2008c).

Mangrove forests occur naturally in intertidal zones along sheltered shorelines and in deltas in tropical regions. They are vital breeding grounds for fish and shrimp and also provide a buffer against coastal hazards such as storms, cyclones, wind and salt spray by reducing wind and wave action (Braatz *et al.* 2007). Evidence collected in Thailand and Sri Lanka following the 2004 Indian Ocean tsunami showed that the loss of life was lower where mangrove and other coastal forests remained intact (Forbes and Broadhead 2008). A contributing factor in the loss of life and property resulting from cyclone Nargis in Myanmar in May 2008 is believed to have been the destruction of mangrove forests, in particular in the Irrawaddy Delta where the area of mangrove forest had been reduced to less than half its level in 1975 (Kinver 2008).

Micro climate

The micro-climate associated with forest areas is often a critical factor in growing cash crops. In East Africa, tea is grown in areas adjacent to montane forests where conditions for tea production are optimal due to constant moisture levels, air temperatures between 10° and 30° C and soil temperatures between 16° and 25° C. The high moisture levels in these montane

⬇ Tea production area and forest distribution in Kenya

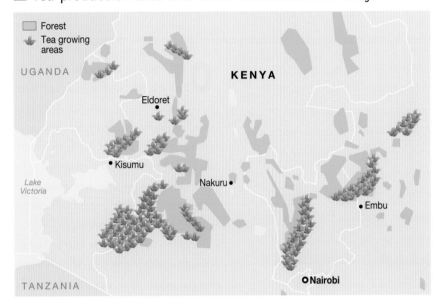

⬇ Forests regulate groundwater level

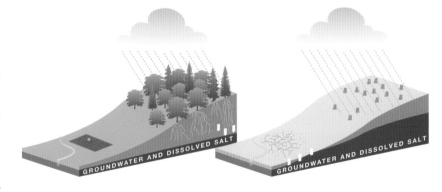

forests combined with the high heat capacity of water reduces the day/night temperature range and lowers the chances of frost.

In Indonesia, areas close to the forests in the Kerinci-Seblat National Park in Sumatra are among the best places in the world to grow cinnamon due both to the relatively cold climate and rich local volcanic soils (World Bank 2007).

These examples help to illustrate how humans and production cycles depend on regulatory services pro-

vided by forests. They also highlight the importance of securing these services – for too long taken for granted. Methods of economic valuation are progressively enabling the incorporating of these services into cost-benefits analysis and helping to make forest conservation not only economically feasible but also desirable.

➡ See also pages 56 & 60

Climate change and its impact on forests - will forests migrate?

Forests play a key role in maintaining a wide range of delicate relationships with nature and its ecosystems. Impacts on the well being of forests likely to be caused by climate change will therefore have a dramatic effect. For example, according to the latest projections, changes in climate will mean that by 2050 the world's ecosystems, including its all important forests, will be releasing more carbon than they are capable of absorbing

Over the last 30 years the world has experienced significant temperature increases, particularly in the northern hemisphere. Meanwhile more climate variability is predicted, with increased precipitation in some areas and extreme dry and hot periods in other regions. Such events will have a substantial effect on forests.

Rising temperatures force many living organisms to migrate to cooler areas, while new organisms arrive. Such movements involve all species, including plants. Some species will seek higher altitudes, others will move further polewards. In temperate regions, plant and tree species can migrate naturally by 25 to 40 kilometres a century. However if, for example, there was a 3°C increase in temperature over a hundred year period in a par-

ticular region, the conditions in that area would undergo dramatic change, equivalent in ecological terms to a shift of several hundred kilometres (Jouzel and Debroise 2007).

In the last few decades scientists have observed the first signs of this process taking place in the northern hemisphere caused, it seems, by temperature rises linked to climate change.

Various studies have noted that a number of bird, tree, scrub and herb species have migrated by an average of six kilometres every ten years, or have sought higher altitudes of between one and four metres (Parmesan 2003). Botanists have also noted that many trees and plants in the northern hemisphere tend to flower increasingly early – on average advancing by two days every ten years – thereby increas-

⬇ Impact of climate change on mountain vegetation zones

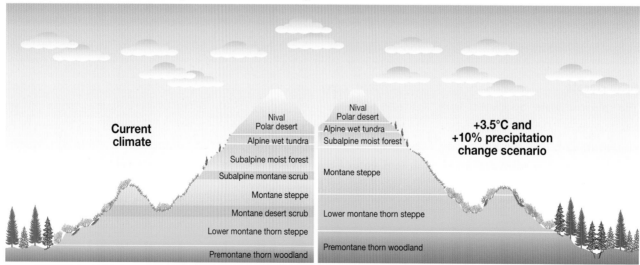

Sources: UNEP/GRID-Arendal 2008; Benitson 1994; Watson *et al.* 1995.

Estimated loss of plant species, 2000-2050

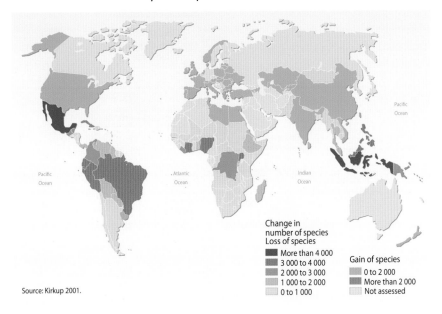

Change in
number of species
Loss of species
- More than 4 000
- 3 000 to 4 000
- 2 000 to 3 000
- 1 000 to 2 000
- 0 to 1 000

Gain of species
- 0 to 2 000
- More than 2 000
- Not assessed

Source: Kirkup 2001.

Carbon stocks trends and projections compared to 1860

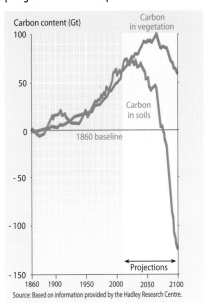

Source: Based on information provided by the Hadley Research Centre.

---

ing the risk of buds being killed by late frost.

Slightly higher temperatures and a greater accumulation of CO_2 in the atmosphere accelerate growth rates of species in forest ecosystems (WRI 2008b). It is estimated that forests in temperate regions have seen a 15 per cent productivity gain since the beginning of the 20th century (Medlyn et al. 2000). In addition, CO_2 fertilization plus increased nitrogen levels and more soil moisture, have all contributed to greater forest productivity over the last century.

Loss of species

Paradoxically, while increased CO_2 levels and other factors have led to forest growth in some regions, the present environmental situation – heavily influenced by climate change – could lead to a massive destruction of forests and the extinction of countless species. For example, modelling focusing on the Amazon region has indicated that 43 per cent of 193 representative plant species could become non-viable by the year 2095 due to the fact that changes in climate will have fundamentally altered the composition of species habitats (Miles et al. 2004).

Global warming is likely to increase the extent of forest fires, as occurred recently in Russia, southern Europe and California. A recent study of various forest conditions in Russia suggests that a 2°C rise in temperature could increase the area affected by forest fires by a factor of between one and a half and two (Mollicone et al. 2006).

Climate variability may also cause plant productivity to drop. During the 2003 heat wave in Europe, there was a 30 per cent fall in plant productivity in continental Europe as a whole. On the other hand mild winters mean that there will be more pests and diseases. Following a series of mild winters in Canada from the 1960s onwards, growing numbers of pine beetles recently caused severe damage to more than 13 million hectares of forest (Brown 2008).

The direct physical effects on forests caused by climate change, such as droughts, storms, fires and insect infestations, could also hurt the productivity of managed forests (WRI 2008b). Both the supply of and demand for forests products will be affected by climate change related events, as will the lives of millions of people – many of them very poor – who are often wholly dependent on forests and associated resources for survival.

→ See also pages 48

Estimated loss of rainfall in Amazonia in the next century

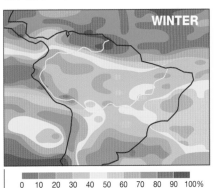

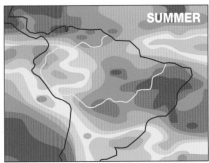

0 10 20 30 40 50 60 70 80 90 100%
Percentage less
Source: Malhi et al. 2008.

The synthesis of 23 climate models shows a decline in rainfall between 1980-1999 and 2080-2099 under mid-range (A1B) global greenhouse gas emissions scenarios. The dry season rainfall is particularly important (winter in north and summer in central and southern Amazonia).

↓ Historical forest carbon balance, 1855-1995

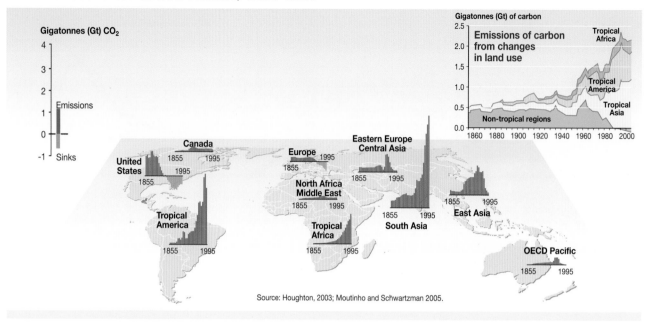

Source: Houghton, 2003; Moutinho and Schwartzman 2005.

Forests and the carbon cycle

*Forests play a vital role
in the global carbon cycle.
Forests absorb carbon through
photosynthesis and sequester
it as biomass, thus creating
a natural storage of carbon*

Through processes of respiration and through the decay of organic matter or burning of biomass, forests release carbon. A carbon 'sink' is formed in the forest when the uptake of carbon is higher than the release.

Carbon stocks in forest areas comprise carbon in living and dead organic matter both above and below ground including trees, the understorey, dead wood, litter and soil. On a global scale, vegetation and soils are estimated to trap 2.6 gigatonnes (Gt) of carbon annually. Yet there are still many uncertainties about the workings of the carbon cycle: the Intergovernmental Panel on Climate Change (IPCC) estimates that the amount of carbon absorbed in the soil and vegetation amounts to anything between 0.9 and 4.3 Gt annually.

Carbon stocks in land based ecosystems are distributed irregularly between tropical and northern latitudes but are mostly concentrated in forest ecosystems and wetlands. Recent research suggests tropical forests play an even more important role in absorbing carbon than previously thought, taking up 1 Gt of carbon every year, or about 40 per cent of the total for land based absorption (Britton *et al.* 2007).

The conversion of forested to non-forested areas in developing countries has had a significant impact on the accumulation of greenhouse gases in the atmosphere, as has forest degradation caused by over-exploitation of forests for timber and woodfuel and intense grazing that can reduce forest regeneration.

Scientists are continuing to investi-

↓ Forest carbon stock per region

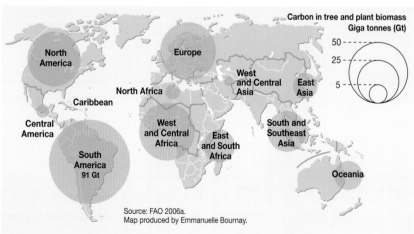

Source: FAO 2006a.
Map produced by Emmanuelle Bournay.

⬇ The carbon cycle

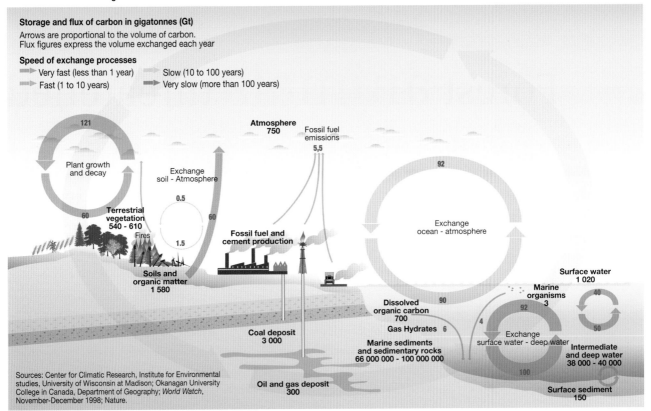

Storage and flux of carbon in gigatonnes (Gt)
Arrows are proportional to the volume of carbon.
Flux figures express the volume exchanged each year

Speed of exchange processes
➤ Very fast (less than 1 year)   ➤ Slow (10 to 100 years)
➤ Fast (1 to 10 years)   ➤ Very slow (more than 100 years)

121

Plant growth and decay

Atmosphere 750

Fossil fuel emissions 5,5

92

Exchange soil - Atmosphere

0.5

Terrestrial vegetation 540 - 610

60

60

Fires

1.5

Exchange ocean - atmosphere

Fossil fuel and cement production

Soils and organic matter 1 580

Surface water 1 020

Marine organisms 3

40

Dissolved organic carbon 700

90

50

Coal deposit 3 000

Gas Hydrates 6

92

4

Exchange surface water - deep water

Marine sediments and sedimentary rocks 66 000 000 - 100 000 000

Intermediate and deep water 38 000 - 40 000

100

Oil and gas deposit 300

Surface sediment 150

Sources: Center for Climatic Research, Institute for Environmental studies, University of Wisconsin at Madison; Okanagan University College in Canada, Department of Geography; *World Watch*, November-December 1998; Nature.

---

gate just how much carbon is emitted as a result of deforestation and forest degradation. The most vital issue is to estimate the true level of global deforestation and forest degradation and the resulting release of carbon stock from the biomass and the soil. In its 4th Assessment Report of 2007 the IPCC said carbon emissions as a result of land-use change – mainly due to deforestation in the tropics – were running at 1.6 Gt of carbon per year in the 1990s, or around 17.4 per cent of the world's total anthropogenic (man-made) emissions of greenhouse gases. However this figure represents only the mid-range estimate, with the IPCC using a range of between 0.5 to 2.7 Gt per year.

A particularly serious impact of deforestation on global climate change is the destruction of forest areas located on peat bogs. Peat areas in tropical zones such as Indonesia and Malaysia only cover about 40 million hectares. Yet when cleared, the destruction of the forest, plus the draining of carbon rich

peat land, results in a massive release of CO_2: it is calculated that such activities now release about 0.5 Gt of CO_2 a year, or 8 per cent of total annual anthropogenic emissions.

In the boreal zone, there are vast expanses of forests on bogs and peat land. The loss of surface permafrost in these areas due to rising temperature will increase the net carbon storage due to vegetation growth, but this increase will be offset by methane emissions (WWF 2008).

Under the United Nations Framework Convention on Climate Change, the Kyoto Protocol was adopted in 1997 with the objective of setting targets to reduce greenhouse gases that cause that cause climate change. Dur-

ing the first Kyoto commitment period (2008-2012), tree plantation projects were considered eligible for carbon credits under the Clean Development Mechanism (CDM), whereas sustainable forest management was excluded from the CDM for a number of political, practical and ethical reasons (Griffiths 2007).

Since carbon emissions from deforestation represent close to one fifth of all anthropogenic greenhouse gases, an initiative was created at the Climate Conference in Montreal in 2006 to "Reduce Emissions from Deforestation and Degradation" (REDD). REDD carbon credits are at the moment included only on the voluntary market.

⬇ Breakdown of carbon storage by region

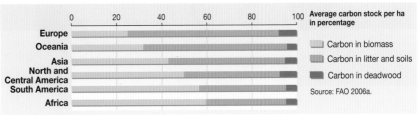

Average carbon stock per ha in percentage

	0	20	40	60	80	100
Europe						
Oceania						
Asia						
North and Central America						
South America						
Africa						

▨ Carbon in biomass
▨ Carbon in litter and soils
▨ Carbon in deadwood

Source: FAO 2006a.

Forest animals threatened

Forests are among the most biologically rich terrestrial ecosystems. Yet deforestation, forest degradation and poaching mean that habitats are lost and the survival of many forest species is increasingly threatened

The World Wildlife Fund (WWF) has identified more than 200 eco-regions as outstanding examples of the diversity of the world's ecosystems; of those, forest regions make up two thirds of the total. Yet while forests contain more than 80 per cent of the world's terrestrial species, the survival of many of them is threatened.

The Convention on Biological Diversity (CBD) estimates that the accelerating rate of deforestation which has taken place over the last century has contributed to reducing the abundance of forest species by more than 30 per cent. The rate of species loss in forest regions is considerably faster than in other ecosystems. Between now and 2050, it is projected that there will be a further 38 per cent loss in abundance of forest species (UNEP-GLOBIO 2008).

The conversion of forests for agricultural use and plantations, fires, pollution, climate change and invasive species all impact on forest biodivers-

ity. Fragmentation of forests due to road, agriculture and human settlement development also impacts on wildlife by reducing the corridors used to move or migrate. In Indonesia over the period 2001-2007, forest clearance, including illegal logging, was found to be taking place in 37 out of 41 national parks, threatening many species and driving the orang-utan towards extinction (UNEP 2006). The decline of the orang-utan and the destruction of its habitat has reached such a level that wildlife conservationists have set up so-called 'orang-utan refugee camps' in certain areas.

Logging and agricultural expansion is also a major threat to amphibians and reptiles. In Haiti, a mega-diverse Caribbean country, amphibian species occur on high mountain ranges and down to sea-level mangrove swamps. The Massif de la Hotte in Haiti alone is home to 32 frog species. However, the future for many of these species is grim, with natural habitats being destroyed for fuelwood and charcoal production. Currently, over 90 per cent of the am phibian species in Haiti are threatened or extinct (Young *et al.* 2004).

Poaching and trade

Animals living in the forest are also at risk from poaching and bush-meat hunting (UNEP 2002, 2005; Dobson and Lynes 2008). In Africa, the bush-meat consumption per capita is higher in logging and mining areas, as the workers are often better off and able to afford bush-meat. Networks of logging roads and tracks also provide hunters with easier access to abundant wild-

⬇ Number of tree species per country in the world

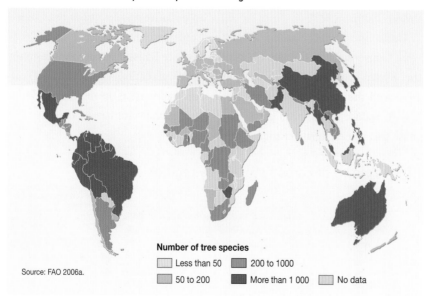

Source: FAO 2006a.

Number of tree species

- Less than 50
- 50 to 200
- 200 to 1000
- More than 1 000
- No data

by habitat loss and poaching

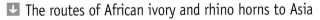

⬇ The routes of African ivory and rhino horns to Asia

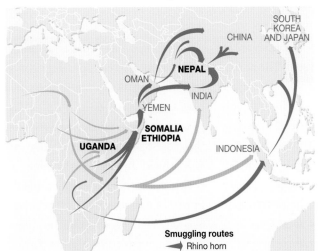

Source: Field investigation, Christian Nellemann, UNEP-GRID/Arendal.

⬇ Wildlife smuggling to, in and from Nepal

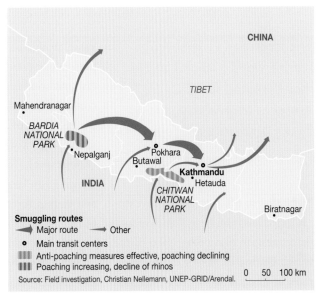

Source: Field investigation, Christian Nellemann, UNEP-GRID/Arendal.

life areas. As a result commercial and subsistence hunting can quickly reach unsustainable levels, leading to local extinction of the targeted wildlife species. In Central Africa, species in danger not only include the larger mammals, such as elephants, rhinos, great apes and other primates, but also porcupine, cane rat, pangolin, monitor lizard and guinea fowl. Bush-meat hunting and trading has now become big business and is one of the main threats to many of the major species in Africa.

Another species under threat from poaching is the rhinoceros. Rhinoceros horn is used in traditional Asian medicine, believed to reduce fevers and even prevent loss of life. Other parts of the rhino, including the skin and bones, are also used for their supposed medicinal qualities. Demand for rhino horn has increased substantially in recent years.

Of the five species of rhino, three are listed in the IUCN Red List as critically endangered. Poaching is not the only way by which rhino horn finds its way to the market. In certain instances and limited to specific populations, trade in rhino horns derived from hunting of rhinoceros is allowed under CITES. However, recent investigations have shown that hunters are abusing regulations and entering rhino horn into commercial trade involving organized crime, corruption, abuse of diplomatic priviliges and money laundering.

Intelligence gathering, regular monitoring and strict enforcement are effective ways of curtailing both illegal logging and poaching activities in forests. The participation of local communities in these activities can facilitate implementation of laws and regulations and secure sustainability. Customs enforcement also plays a critical role in controlling trade in various species.

⬇ Number of animal species per biome/ecosystem

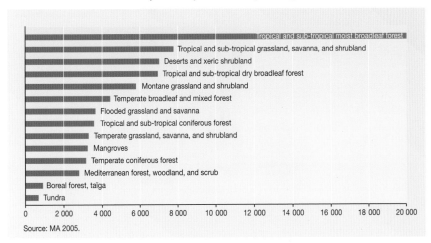

Source: MA 2005.

The forests of Central Africa

The forests of Central Africa are the world's second largest moist tropical forests, exceeded in area only by the Amazon Basin forests

The forests of Central Africa have suffered less from large scale clearance than forests in many other parts of the world. During the 1990s deforestation rates in Central Africa were estimated to be 0.35 per cent per year. In the 2000 to 2005 period, deforestation rates in the region showed marked differences between the countries concerned; while 1 per cent of Cameroon's forests were lost every year during this period, the figure was 0.24 per cent for the Democratic Republic of the Congo (DRC) and 0.05 per cent for Gabon (FAO 2006a).

About 1 000 different tree species have been recorded in the area, with some 100 species having commercial value, though only 40-50 species are at present harvested and sold. The northern and southern parts of the region consist of semi-deciduous forests, with an abundance of commercially valuable species such as the African mahogany (*Khaya* spp.). The central part of the Congo Basin largely consists of gallery and swamp forest. In the eastern part of the region as many as 300 tree species can be found on a single hectare.

Some 30 million people, comprising 150 different ethnic groups live in the forests of Central Africa. The majority are indigenous and include the Pygmies, as well as many groups of Bantu origin that have been forest dwellers for more than 1000 years and have intermixed with Pygmy, Ubangi and Sudanic populations. They are distinguished among themselves by their degree of nomadism (hunter-gatherer) and their dependence on farming, mainly traditional shifting-agriculture. Most communities exist at subsistence levels.

The Congo Basin forests are under threat from deforestation, degradation and fragmentation, even in areas that have not yet been opened up by logging operations. Satellite images show that forest loss and degradation are also caused by the displacement of people due to war and conflict, and the impact of mining activities. The images also highlight destruction in previously untouched forest areas particularly along the tributaries of the Congo River, most likely related to population growth and settlement. In the densely populated mountainous regions and high plateaus of western Cameroon and the east of DRC, the shortening of fallow periods, combined with the conversion of abandoned fields into pasture prevent any forest regeneration.

Across the region 76 per cent of forests have been identified as productive or commercially exploitable. By 2000 nearly all productive forests outside protected areas in the central Congo

⬇ Forest conservation and wood production

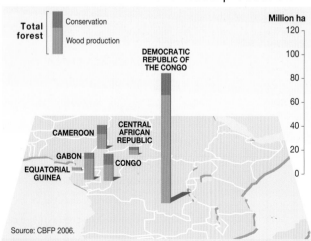

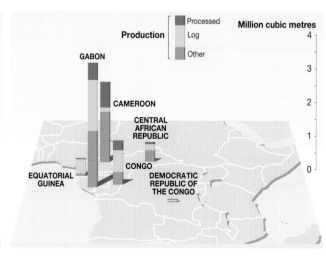

Source: CBFP 2006.

⬇ Increase in forest concessions in Cameroon

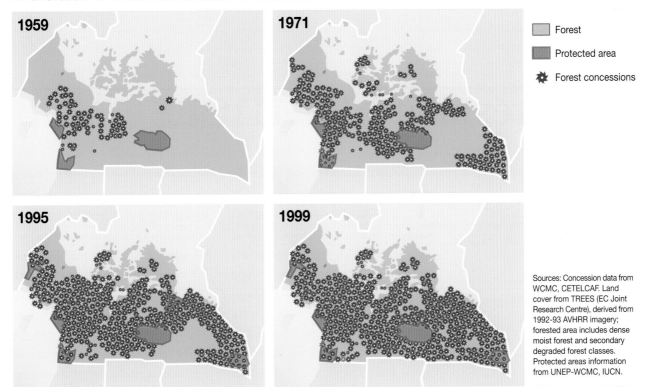

1959

1971

1995

1999

Forest

Protected area

✳ Forest concessions

Sources: Concession data from WCMC, CETELCAF. Land cover from TREES (EC Joint Research Centre), derived from 1992-93 AVHRR imagery; forested area includes dense moist forest and secondary degraded forest classes. Protected areas information from UNEP-WCMC, IUCN.

Basin had been allocated to timber concessions. If selective harvesting practices are followed, the impacts of the timber operators would theoretically remain relatively limited. However the implementation of such practices has been hampered by the lack of a regulatory framework and control as well as attention to the dynamics of the forest. Fast growing global demand for timber has increased the risk of unsustainable forestry practices, illegal logging and corruption. It is now of vital importance to introduce and regulate sustainable forestry practice and as well as independent certification. Equally important is to strengthen the rights of forest based peoples and communities.

In the DRC logging contracts covering 25 million of the 41 million hectares originally allocated were cancelled in 2002. At the same time, a moratorium on the further allocation of forest concessions was put in place. An inter-ministerial commission is at present verifying the legality of titles of all 156 forestry concessions. Meanwhile two of the largest operators of logging concessions in the DRC are likely to have their concessions certified in the near future.

The establishment of protected areas is a crucial element for the long-term sustainability of contiguous forests. In 1999, the governments of Central Africa agreed on the Yaoundé Declaration for the conservation and rational use of Central African forests. As a result, new protected areas have been established in Equatorial Guinea (10), Gabon (13) and Cameroon (4).

The initiative on Reduced Emissions from Deforestation and Degradation (REDD) presents new opportunities to conserve forests. If countries in central Africa are to benefit from a REDD mechanism, national incentive schemes geared towards poor communities must address the complex problem of forest destruction by slash-and-burn farmers without land title.

⬇ Logging concessions and protected areas

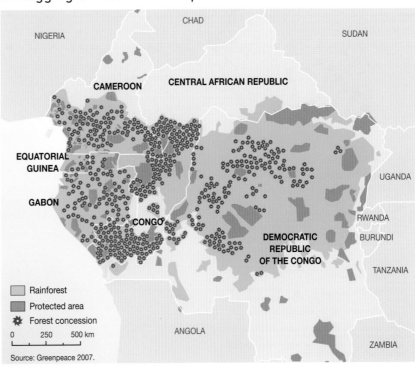

Rainforest

Protected area

✳ Forest concession

0   250   500 km

Source: Greenpeace 2007.

The forests of Southeast Asia

Southeast Asia, containing the world's third largest tropical forests, is experiencing deforestation rates higher than almost anywhere else on Earth. The region's forests are endangered by conversion to agriculture or other land uses, such as oil palm plantations, logging (both legal and illegal) and climate change.

Tropical rainforests cover approximately 60 per cent of the region's total forest area, with tropical moist deciduous forests and tropical dry forests each accounting for around 15 per cent and mountain forests another 10 per cent (FAO 2001b). Mangrove forests, found in the interface between land and sea, represent about one third of the world's total mangrove cover (FAO 2007). Fresh-water and peat swamp forests are also present. Dry forests include deciduous dipterocarp forests and mixed deciduous woodlands often containing some very valuable tree species, such as teak *(Tectona grandis)*, and trees from the dipterocarp family (Dipterocarpaceae). In Malaysia, Indonesia and the Philippines the montane (evergreen) rainforest, most developed at altitudes between 1 400 and 2 400 m, still covers relatively large areas.

Almost the whole of Southeast Asia was covered by forest 8 000 years ago (Billington *et al.* 1996). Today only about half the land area is covered by forest and most of the countries in the region have experienced rapid declines in forest area. It is calculated that the region is losing about 1.2 per cent of its remaining forest area each year, with Cambodia, Indonesia and the Philippines reporting annual losses of two per cent over the last five years (FAO 2006a).

Conversion to agriculture, including the recent expansion in the area devoted to oil palm plantations, continues to be the main cause of forest loss in the region. Meanwhile, a large portion of mangrove areas has been converted to shrimp farms or rice cultivation.

Logging and pulpwood clear-cutting have also been major causes of deforestation in some areas. The high proportion of valuable timber species in the lowland forests and easy access to the coast and shipping routes are

⬇ Forest and biodiversity under threat by economic development

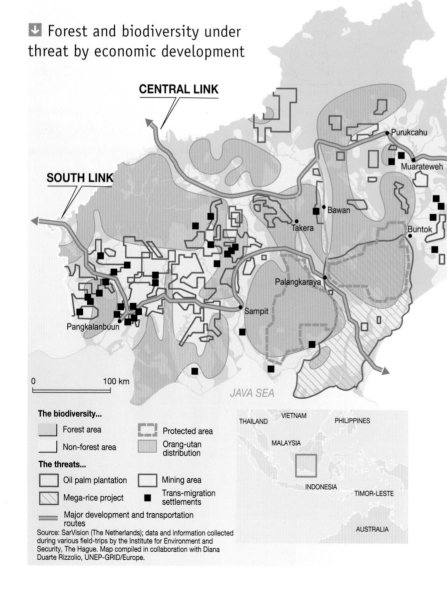

The biodiversity...
- ▨ Forest area
- ☐ Non-forest area

The threats...
- ☐ Oil palm plantation
- ▨ Mega-rice project
- ═ Major development and transportation routes
- ⬚ Protected area
- ▨ Orang-utan distribution
- ☐ Mining area
- ■ Trans-migration settlements

Source: SarVision (The Netherlands); data and information collected during various field-trips by the Institute for Environment and Security, The Hague. Map compiled in collaboration with Diana Duarte Rizzolio, UNEP-GRID/Europe.

⬇ Future expansion of palm oil in Indonesia

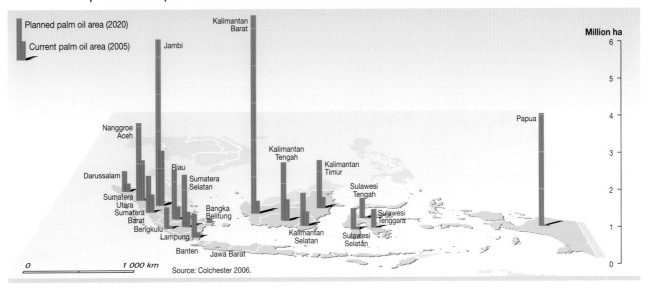

Planned palm oil area (2020)

Current palm oil area (2005)

Million ha
6
5
4
3
2
1
0

Kalimantan Barat
Jambi
Papua
Nanggroe Aceh
Kalimantan Tengah
Kalimantan Timur
Darussalam
Riau
Sulawesi Tengah
Sumatera Selatan
Sumatera Utara
Sumatera Barat
Bangka Belitung
Sulawesi Tenggara
Bengkulu
Lampung
Kalimantan Selatan
Sulawesi Selatan
Banten
Jawa Barat

0   1 000 km

Source: Colchester 2006.

---

among the reasons for this. Most of the more accessible forests in the region have been logged at least once. Commercial logging in Papua New Guinea, for example, has been heavily concentrated in forest areas that are accessible by bulldozers, trucks and coastal shipping.

Recent studies conclude that by 2021 approximately 80 per cent of the commercially accessible forests which were present in Papua New Guinea in 1972 would have been cleared, commercially logged or affected by low intensity fires (UPNG Remote Sensing Centre 2008). Even forests in protected areas in Kalimantan, Indonesia, are being logged and have declined by more than 56 per cent between 1985 and 2001 (Curran et al. 2004).

The interests of commercial activities such as palm oil plantations or mining, often clash with the interests of communities, small farmers and indigenous people when it comes to management of natural resources. Land rights issues are often at the heart of such conflicts.

For example, in Indonesia there is growing evidence of human right violations associated with the palm oil industry (Friends of the Earth 2008). Indonesia has set ambitous targets for oil palm expansion and such conflicts are likely to intensify if human rights

issues are not appropriately addressed.

Due to the rapid rate of deforestation and forest degradation in the region, there are growing concerns about increases in greenhouse gas emissions. Of particular concern are the peat swamp forests, where peat deposits are up to 20 metres thick and contain vast reserves of near-surface terrestrial organic carbon. Out of 27 million hectares of peat land in Southeast Asia, an estimated 12 million hectares has been deforested or degraded over the past ten years (Hooijer et al. 2006).

The establishment of conservation areas and better forest management practices are essential tools in the battle to save the tropical forests. One major step was taken in 2007 when the

Forestry Ministers of the three countries involved – Indonesia, Brunei and Malaysia – signed the Heart of Borneo Declaration and 220 000 square kilometres – or an area almost as big as Great Britain – was turned into a large network of protected areas and forest areas managed according to the principles of sustainable forestry.

The EU Action Plan on Forest Law Enforcement, Governance and Trade (FLEGT) is also being seen as a good example of how to develop partnership agreements between the producer and the consumer countries to combat illegal timber trading.

➡ See also pages 20, 24, 30

⬇ Colonization of West Papua, Indonesia

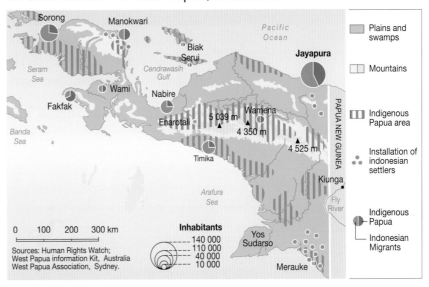

Sorong
Manokwari
Pacific Ocean
Biak
Serui
Jayapura
Seram Sea
Cendrawasih Gulf
Wami
Nabire
Fakfak
Enarotali   5 039 m   Wamena
Banda Sea
4 350 m
Timika
4 525 m
Arafura Sea
Kiunga
Fly River
Yos Sudarso
Merauke

PAPUA NEW GUINEA

Plains and swamps

Mountains

Indigenous Papua area

Installation of indonesian settlers

Indigenous Papua

Indonesian Migrants

0   100   200   300 km

Inhabitants
140 000
110 000
40 000
10 000

Sources: Human Rights Watch;
West Papua information Kit, Australia
West Papua Association, Sydney.

The Amazon, the largest

The Amazon Basin, stretching over Bolivia, Brazil, Colombia, Ecuador, French Guyana, Guyana, Peru, Suriname and Venezuela, contains the world's largest tropical rainforest, and is home to more than 30 million people. It is also an ecosystem with unparalleled rich biodiversity.

The dominant ecozone in the Amazon contains tropical rainforest, tropical moist deciduous forest, tropical mountain forest and tropical dry forest. The wettest type of vegetation is found in the upper basin of the Amazon River, in the State of Amapà in Brazil and on the west coast of Colombia where luxuriant, multilayered evergreen forest grows to heights of up to 50 metres tall (FAO 2001b).

The more extensive rainforest vegetation is somewhat drier and occurs in the Amazon Basin and on the eastern foothills of the central Andes. It consists of multilayered forest up to 40 m high, with or without emergent trees, and is mainly evergreen (FAO 2001b). Of the approximately 300 tree species that may be found in a single hectare of rich Amazon rainforest, only 30 to 50 are considered to be of commercial use (Grainger 1993).

The Amazon population is mainly urban, but many indigenous groups live in villages deep in the forest. Brazil has 225 indigenous groups, of which 170 live in the Amazon. Some 460 000 indigenous people, or 0.25 per cent of the Brazilian population, live inside

580 officially recognized indigenous territories, the majority in the Amazon. These territories represent 13 per cent of the national territory, and more than one fifth of the Brazilian Amazon. The states with the largest number of isolated tribes, such as Pará, Mato Grosso and Rondônia, are unfortunately also under growing pressure from deforestation and together, they accounted for about 85 per cent of Brazil's total deforestation over the 2006-2007 period (INPE 2008).

Over the past 40 years, about a fifth of Brazil's Amazon rainforest has been deforested (Reuters 2008). Official statistics show that annual deforestation has been close to 20 000 square kilometres over the last 10 years, reaching a peak of 27 429 square kilometers in 2004, and then being reduced annually to 11 224 square kilometers in 2007 (INPE 2008).

The most obvious reason for deforestation is the conversion of forest lands for cattle ranching and agricultural crops, industrial activities and logging for timber. Transportation infrastructure has been linked to aggressive and rapid change in land use, with new roads

⬇ Deforestation causes in Brazil

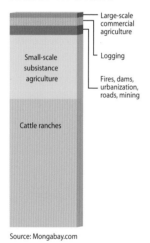

Source: Mongabay.com

⬇ Socio-economic indicators for the Legal Amazon region in 2004

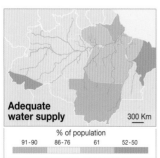

Adequate water supply
300 Km
% of population
91-90   86-76   61   52-50

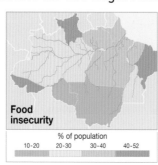

Food insecurity
% of population
10-20   20-30   30-40   40-52

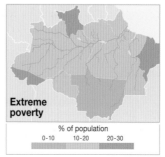

Extreme poverty
% of population
0-10   10-20   20-30

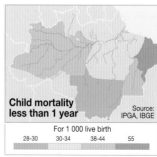

Child mortality less than 1 year
Source: IPGA, IBGE
For 1 000 live birth
28-30   30-34   38-44   55

rainforest in the world

⬇ 2050: Worst case scenario for the Amazon rainforest

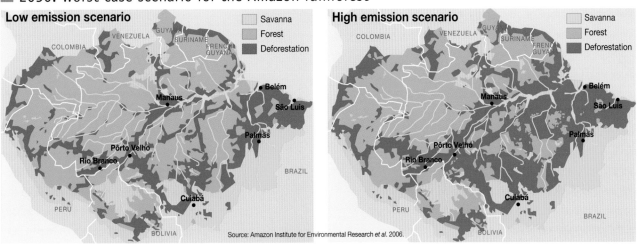

Low emission scenario

Savanna
Forest
Deforestation

High emission scenario

Savanna
Forest
Deforestation

Source: Amazon Institute for Environmental Research *et al.* 2006.

making previously remote areas of forest accessible to farmers and ranchers, thus facilitating conversion of forest land to agricultural crops and use as pasture. It is estimated that 90 per cent of deforestation in the Amazon region has occurred within 50 kilometres of roads (INPE, 2008). There are indications that after a relative drop in deforestation over the 2004 to 2007 period, deforestation rates in Brazil have again been rising encouraged by record world prices for both soya beans and beef, pushing the agricultural frontier ever further into the rainforest.

Global climate change has already contributed to rising temperatures in the Amazon which, when combined with deforestation, have led to a cycle of lower precipitation and a greater frequency of droughts. Researchers at Brazil's National Institute for Space Research say that the Amazon could reach a tipping point – the point at which deforestation and climate change combine to trigger self-sustaining desertification –

in 50-60 years (Reuters 2008).

The development of government policies such as *Zona Franca Verde* or green free trade zones, and the 80:20 land use practice have led to some decline in deforestation. Further to refrain the deforestation, the Brazilian Government was one of the first to adopt a system of payment for environmental services.

In 2008, the Brazilian Government announced the enlargement of a network of protected areas under the Amazon Region Protected Area Programme (ARPA), to cover nearly 600 000 square kilometres of the Amazon by 2016. The government has also announced the creation of a US$21 billion fund called the Amazon Fund, to pay for projects designed to prevent deforestation, support conservation and sustainable development of the Amazon region (AFP 2008).

➡ See also pages 10, 20, 52, 58

⬇ Deforestation in Brazil compared with the area of Turkey

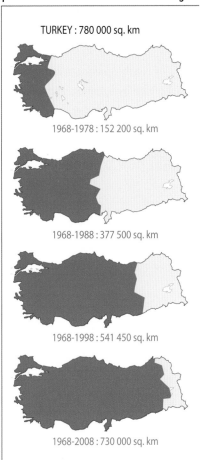

TURKEY : 780 000 sq. km

1968-1978 : 152 200 sq. km

1968-1988 : 377 500 sq. km

1968-1998 : 541 450 sq. km

1968-2008 : 730 000 sq. km

The boreal forests

Boreal forests and taiga in extreme northern areas of the Arctic cover some 386 million square kilometers representing one of the largest forest biomes in the world (Olson 2001).

Boreal forests generally have a low level of species diversity. The main tree species are spruce (*Picea* spp.), pine (*Pinus* spp.), larch (*Larix* spp.) and fir (*Abies* spp.), which are all coniferous, but a few deciduous species such as birch (*Betula* spp.), aspen and poplar (*Populus* spp.) also occur, especially in the early stages of forest succession.

Forest fires are a central element involved in the natural succession of boreal forests while small-scale dynamics include storms, insects and pathogenic fungi, in particular on moderately moist sites (Väisänen 1996). In the intact boreal forests of North America, forest fires are mostly caused by lightning. However, the latest research suggests that fires caused by humans play a larger part in the boreal for-

The dynamics of boreal forest ecosystems provide the basis for conservation and sustainable use of forest resources, yet the debate continues concerning clear-cutting practices and, more recently, on the impacts of climate change

est of Eurasia than earlier assumed (Achard *et al.* 2008; Ruckstuhl *et al.* 2008).

Clear-cutting as a forestry practice has been much discussed in the context of boreal forests. It is seen as essential not only from a commercial perspective but also from a biodiversity point of view: the effects of clear-cutting are comparable with natural fire dynamics, resulting in patches of even-aged homogenous forests, typi-

cal of boreal forests. Since the late 1960s, debate has focused on the maximum size of these individual clear-cuts, especially after large scale cutting areas were introduced in order to increase the economic efficiency of boreal forestry. More recently, the average clear-cut size has been decreasing so as to better adapt to the natural dynamics of the ecosystem. In Europe, this is seen as a positive trend since a high degree of variation is beneficial to the fauna and flora and increases the resilience of Europe's mostly semi-natural forest landscapes to pests and diseases. For example, in the 1990s, about 94 per cent of clear-cuts in Halsingland, Sweden, were smaller than 100 hectares, as compared to about 83 per cent in the 1980s (EEA 2006).

In the province of Ontario in Canada, about 80 per cent of all harvested forests in the boreal zone have to be in blocks of less than 260 hectares. Larger clear-cuts are only allowed under strict conditions and must be recorded and approved in forest management plans (MNRO 2008). In the province of British Columbia, the average clear-cut size has also been decreasing – to just under 30 hectares in 1998 (BC Ministry of Environment 2000).

Overall, the area of boreal forest has remained fairly stable between 1990 and 2005. While some countries such as Sweden have experienced a slight forest gain, other countries, including Finland and the Russian Federation, have had a very slight forest loss (FAO 2006a; MCPFE 2007).

According to the Fourth Assessment Report of the Intergovernmental Panel on Climate Change (2007), the boreal zone is likely to be strongly affected by climate

⬇ Main flow of illegally logged timber to Europe

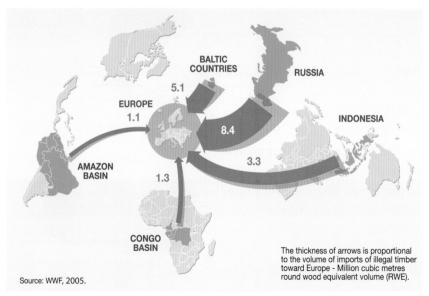

The thickness of arrows is proportional to the volume of imports of illegal timber toward Europe - Million cubic metres round wood equivalent volume (RWE).

Source: WWF, 2005.

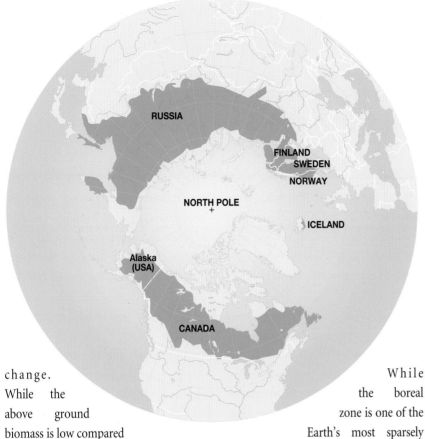

← Boreal forest extent

change.
While the
above ground
biomass is low compared
to forests in temperate or tropical environments, the organic soil content can be very high. This particularly applies to boreal forests in the Russian Federation. A rise in temperature is likely to cause increased organic decomposition which in turn will lead to the accelerated release of substantial amounts of methane into the atmosphere. Analysis also suggests that po-pulations of insects such as mountain pine beetles which periodically erupt into large-scale outbreaks have expanded over the past 40 years (Taylor 2006). As the result of such insect infestations tree mortality rates increase, further reducing forest carbon uptake.

⬇ Cutting smaller parcels in Halsingland, Sweden

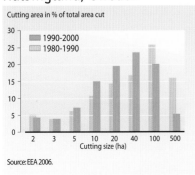

Cutting area in % of total area cut

■ 1990-2000
▩ 1980-1990

Cutting size (ha)

Source: EEA 2006.

While the boreal zone is one of the Earth's most sparsely populated zones it is home to many indigenous peoples. The Russian Federation recognizes more than 40 Small-numbered Indigenous Peoples of Russia living in the boreal and taiga zone, each having a maximum of 50 000 members. The Canadian Census of Population 2006 found that there are over 1.17 million aboriginal people in Canada, representing nearly 4 per cent of the country's total population (Statistics Canada 2008).

Mining for minerals, oil and gas is prevalent in many parts of the boreal zone and often is the cause of environmental concern. In 2008, the Canadian government decided to protect at least 225 000 square kilometers of Ontario's vast boreal forest from mining and other resource development projects and restrict land use to tourism and traditional aboriginal practices such as hunting and fishing (MNRO 2008).

Illegal logging is a serious issue, especially in remote areas of the Russian Federation. Estimates of the extent of illegal logging in the Federation vary considerably, with between 5 and 30 per cent of the boreal forest area affected (MCPFE

2005). Some link the surge in illegal logging in Russia to the adoption of a new forest code which led to changes in issuance of logging permits (Taiga Rescue Network 2008). Official statistics from the Baltic countries estimate the illegal harvesting at 0.26 million m3, just less than 1 per cent of the total cutting volume. However, WWF reports illegal timber exports from the Baltic countries to Europe at 5.1 million m3, an estimate which is 20 times higher than the officially reported illegal harvesting.

On the other hand, the countries in the boreal zone are forerunners in monitoring progress towards sustainable forest management.

⬇ Illegal logging in the Baltic countries

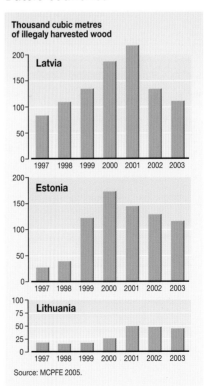

Thousand cubic metres of illegaly harvested wood

Source: MCPFE 2005.

Forests and fires

Often unquantified, the social and economic impacts of forest fires are considerable: lives are lost, health problems occur, animals are killed and the environment suffers

Over the summer of 2007, Greece was hit by its most devastating forest fires in 50 years. People were burned to death. Trees burst into flames like giant matchsticks. For days a dense cloud of smoke blocked out the sun. In all, at least 270 000 hectares of forest, olive groves and farmland were destroyed. Though the scale and ferocity of the Greek fires was unusual – blamed on record high temperatures, a prolonged drought and in some areas the activities of arsonists – such conflagrations regularly devastate forests around the world.

Satellite data from 2000 revealed that 350 million hectares of land was affected by vegetation fires worldwide – an area slightly bigger than the whole of India – much of which was wooded savannah, woodland and forest. According to the data, a large proportion of the burned area was in sub-Saharan Africa.

Each year in the Mediterranean region alone about 50 000 separate fires sweep through up to 1 million hectares of forest and other woodland. In 2007, wildfires ripped through southern California, burning more than 200 000 hectares of trees, destroying homes, and claiming lives. These siege fires are ten times the size of the average forest fire of 20 years ago and are becoming increasingly frequent (Wood 2007). Figures for 2005 indicate that nearly 4 million hectares in the US were burned – more than twice the average annual figure in the previous ten years.

Many of the world's ecosystems have evolved under the influence of fire and need it to regenerate. Up to the early 1990s, forest fires in Brazil were seen as a management tool, used to transform forest biomass into soil nutrients or eliminate invasive species and weeds (Mountinho and Schwatzmann 2005). It is still seen an important factor in ecosystem regeneration. However, in some parts of the world, for example

⬇ Smoke over Southeast Asia on 11 September 1997

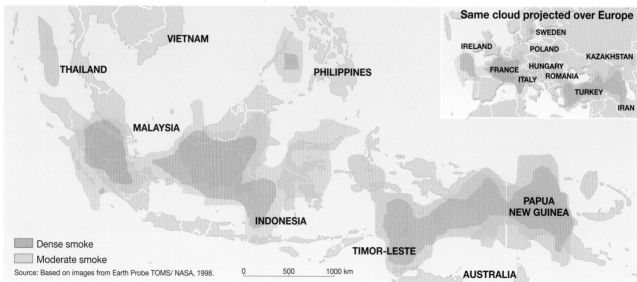

Dense smoke
Moderate smoke
Source: Based on images from Earth Probe TOMS/ NASA, 1998.

0   500   1000 km

⬇ Trends in occurrence of wild fires

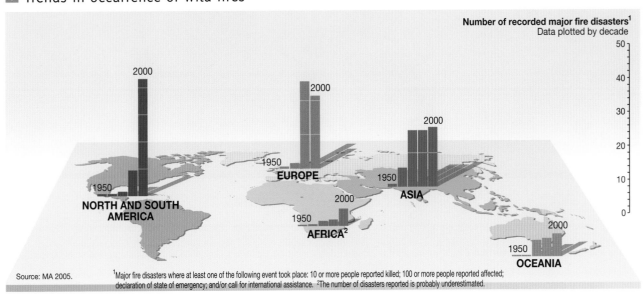

Number of recorded major fire disasters[1]
Data plotted by decade

Source: MA 2005. [1]Major fire disasters where at least one of the following event took place: 10 or more people reported killed; 100 or more people reported affected; declaration of state of emergency; and/or call for international assistance. [2]The number of disasters reported is probably underestimated.

in the Amazon, researchers have begun to warn of a growing forest fire risk (Mountinho and Schwartzmann 2005).

Forest fires also cause widespread damage and are responsible for large emissions of carbon into the atmosphere. In some regions there is evidence of an ever increasing number of fires affecting larger areas and burning with greater severity. Climate change and the lack of sustainable land use policies are contributing factors in this increase. The El-Niňo effect is also a factor, as it contributes to increases in the frequency of drought and lightning strikes.

The majority of fires are caused by agricultural burning to convert forests to ranch or crop land, by careless burning of residues and waste, by burning to improve hunting and by arson. In many regions, the spread of urban development is also associated with increased incidence of fires.

As deforested areas expand, changes in the landscape and micro climate occur: the forest floor dries up which, in turn, makes it more vulnerable to fire. Ecosystems can change as a result of fires. Fires in the understorey of humid rainforests can cause tree mortality and canopy openness. Frequent fires and land over-use means forests are increasingly impoverished and, in some cases, such land becomes savannah (Mountinho and Schwartzmann 2005).

In remote areas of Canada and Russia lightning is a major cause of fires. The remoteness of these areas often results in fires developing into serious conflagrations.

Most countries have extensive fire regulations but there is often a lack of regulatory enforcement and in some countries policies and laws are contradictory. The Association of Southeast Asian Nations (ASEAN) introduced a zero burning policy for the region in 1999 but this has been largely ineffective: it has been found that the use of fire, mostly for agricultural purposes, is a far more important factor in maintaining rural livelihoods than previously thought.

→ See also pages 20, 34, 36

⬇ Estimate of area of vegetation destroyed annually by fire by region

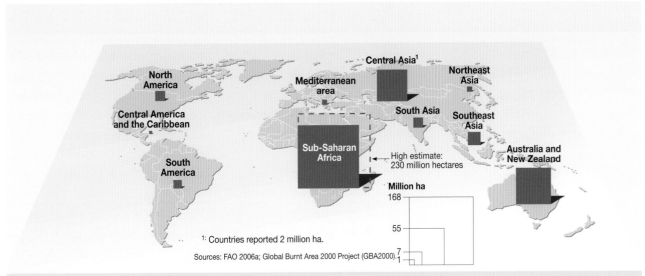

[1]: Countries reported 2 million ha.

Sources: FAO 2006a; Global Burnt Area 2000 Project (GBA2000).

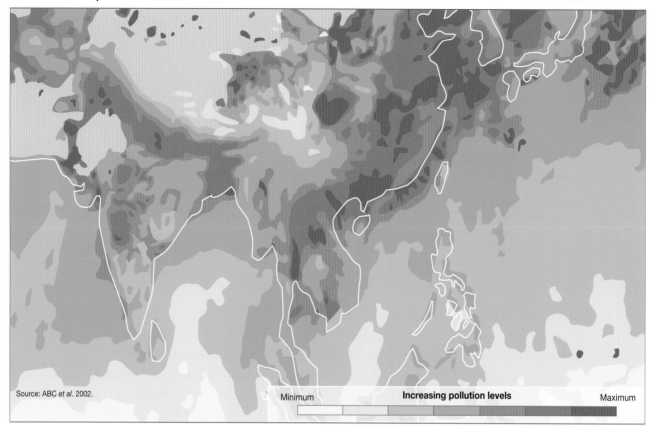

Source: ABC *et al.* 2002.

Minimum   **Increasing pollution levels**   Maximum

Forests suffer from air pollution

Air pollution has had an impact on trees and forests since the early days of the industrial revolution. In some cases the damage has been minor but in others entire forests have been killed. Though the processes causing such damage are well understood, knowledge is generally restricted to a few well-known tree species and lichens. Less understood is the impact of air pollution on birds or animals and its long term effects on ecosystems in general

Air pollution takes many forms. In the past sulphur dioxide pollution – the result of burning high sulphur coal in the factories of industrialized nations – was widespread. Though such pollution has been sharply reduced in much of the western world, it is still a significant pollutant in many fast developing countries, particularly China and India. Despite stringent air quality regulations in North America, there are areas where sulphur dioxide is a continuing problem and the cause of widespread damage to trees and forests.

The Province of British Columbia, Canada, is now a major source of natural gas, with the number of gas wells rapidly increasing. Many of these wells have quantities of 'sour gas' – gas with significant amounts of hydrogen

sulphide. If gas is flared efficiently, few pollutants escape other than carbon dioxide. However it seems that in most cases a wide range of pollutants is produced.

These chemical cocktails damage nearby areas. Lichens appear particularly sensitive but damage to trees has also been recorded. While emissions from a single site might have little overall impact, problems are likely to build up as many thousands of wells are sunk (Case, 1980). Extensive forest damage could also occur as a result of the present exploitation of tar sand oil in the province of Alberta, Canada.

The most direct effect of air pollution on forests occurs when trees are exposed to a particular pollutant and, as a result, suffer damage to their foliage. This generally involves the uptake of gaseous pollutants through the stomata – the area where damage is often first apparent. The symptoms vary according to species and pollutants. The vulnerability of trees and forests to other pests, such as invasions by bark beetle, can also increase as a result of chemical pollution.

Diagnosis of problems affecting trees can be complicated by the presence of similar symptoms caused by other stresses besides air pollutants. For example forests and trees quickly react to drought, with discoloration, loss of foliage and die back ultimately leading to tree mortality. As the frequency of climate change induced drought is expected to increase in some areas, the health of more forests is likely to be at risk.

In the 1980s, acidic deposition – commonly termed acid rain – was considered to be a threat to ecosystems in Europe and North America. Extensive surveys indicated the condition of forests across these regions was rapidly deteriorating. Although air quality in Europe has improved considerably over the years, trees are still under stress (MCPFE 2007).

More recently, ozone – formed through the interaction of nitrogen oxides, volatile organic carbons (VOCs) and sunlight – has emerged as a significant problem in many regions. With its characteristic symptoms, such as purplish-brown stippling on the upper surfaces of leaves, ozone damage is particularly associated with forests near large urban sites such as Mexico City and Los Angeles. Trees and forests have also been affected in many other regions, including much of south-central Europe, and damage is sometimes associated with the highly toxic peroxyacetyl nitrate (PAN).

Unlike sulphur dioxide, sourced mainly from large scale operations such as coal-fired power stations, nitrogen oxides are emitted from multiple sources, with vehicle exhausts a big factor. Such emissions – which can also adversely impact on trees and forests – are more difficult to control.

Some of the earliest and most detailed studies of the effect of air pollution on forests were conducted around actual sources of pollution, such as smelters (Kozlov and Barcan 2000). The studies showed pollution levels were severe enough to kill all trees and vegetation in the vicinity. Such deserts – devoid of all nature – still surround several smelters in Russia. As soils are usually heavily polluted by a range of chemicals, any clean-up is extremely difficult.

While the type of direct injury to forests seen near smelters in North America and in lignite-burning areas of eastern Europe is unlikely to continue for much longer due to tighter emission regulations, air pollution-related damage to trees – much of it in the form of ozone – is likely to persist in these areas into the foreseeable future. Outside these regions, particularly in Asia and South America, damage caused by air pollution seems likely to increase.

Recent brown cloud

One of today's most urgent and widespread air pollution problems is the smog that extends over much of China, with both urban and rural air quality often extremely poor. At times visibility over a wide area is less than 100 metres with a thick pall of smog clouding the sky.

The impact of smog on vegetation in China is not well-known, though it is clear widespread damage is being done to trees, forests and other vegetation around sites of particularly high pollution. A similar situation pertains in India.

⬇ The Kola Peninsula under threat from deadly emissions

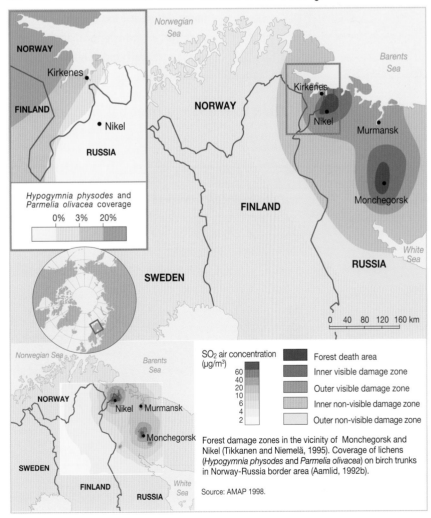

Forest damage zones in the vicinity of Monchegorsk and Nikel (Tikkanen and Niemelä, 1995). Coverage of lichens (*Hypogymnia physodes* and *Parmelia olivacea*) on birch trunks in Norway-Russia border area (Aamlid, 1992b).

Source: AMAP 1998.

Local forest management

⬇ Indigenous land in Amazonia

For generations, local communities around the world have relied on forests not only for their livelihoods, but also as an integral element in their cultural, spiritual and social systems. At present some of the world's most successful forest stewardship schemes exist where communities have either title to their forestlands or have primary rights to use and manage the land. The lesson is clear: when local people have a vested interest in the land, forests and the communities in and around them sustain each other.

Local forest management takes many different forms – from co-management, in which local and central officials share responsibilities, to participatory management or community-based management, in which the central government devolves power to the local level (CIFOR 2008). Community forest management has emerged as the dominant approach in developing countries, often due to management failures of central governments (Pandit *et al.* 2008). The proportion of forests owned or administered by local communities has doubled in the past 15 years (Scherr *et al.* 2003). In 18 developing countries with the largest amounts of forest cover, more than 20 per cent of forests are owned, managed, or reserved for communities (Molnar *et al.* 2003).

In Tanzania, for example, more than 90 per cent of the population uses firewood for domestic energy. At the same

⬇ Trends in deforestation in the Xingu river basin, Brazil

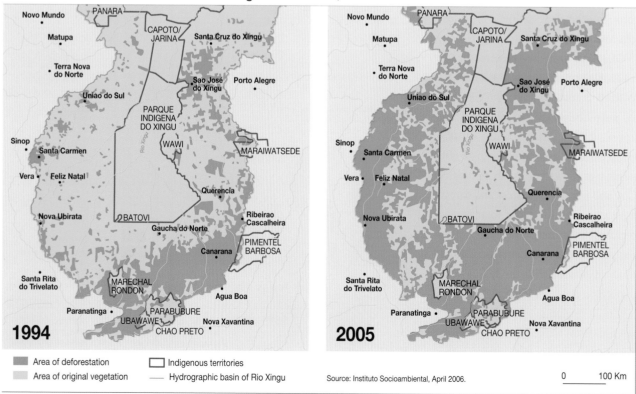

1994

2005

| Area of deforestation | Indigenous territories |
| Area of original vegetation | Hydrographic basin of Rio Xingu |

Source: Instituto Socioambiental, April 2006.

0   100 Km

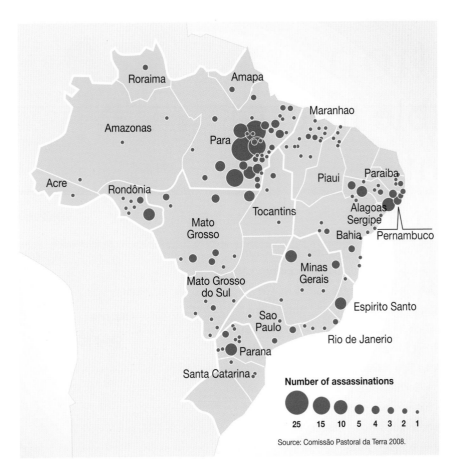

⬅ Loss of human lives
due to conflict over land
1997-2007, Brazil

Number of assassinations

25 15 10 5 4 3 2 1

Source: Comissão Pastoral da Terra 2008.

resistant to ceding control of forests to local people.

The legal right to manage the land is not always a sufficient safeguard, particularly when the rule of law is not upheld. With the doubling of the world's population since 1960, land development pressures have mounted. The rights of forest communities have too often been trampled on as prejudice and competing interests have led to intimidation and violence against indigenous and traditional communities.

time, the country's unprotected forest areas have come under increased pressure from human settlements, illegal harvesting of wood, fires and mining. The Tanzanian government recognized the need to take action in the 1990s and adopted forest and rural development policies to encourage local forest management. Participatory forest management is now operating or being established in more than 1 800 villages, encompassing more than 3.6 million hectares of forest land – equal to approximately 11 per cent of the country's total forest cover.

In Mexico, ownership/user rights to forest resources are mainly allocated to communities that manage and use the forests. About 60 per cent of the total forest area in the country is common property land owned by *ejidos*, groups of people who have the right to use the land, or by indigenous communities (FAO 2009).

In Brazil, the Government has demarcated over 105 million hectares of indigenous territories to establish

488 indigenous reserves. These reserves represent 12.5 per cent of Brazil total area. In so doing indigenous communities have been granted use of the lands forever (Povos Indigenas 2008).

At the international level, the concept of local resource management has been recognized through the idea of Community Conserved Areas (CCAs). The 7th Conference of Parties to the Convention on Biological Diversity acknowledged the CCA concept and called for *"full and effective participation by 2008 of indigenous and local communities… in the management of existing, and the establishment and management of new, protected areas"* (CBD 2004).

The goal of "full and effective participation" in the management and creation of protected areas has yet to be fulfilled. In some cases, the creation of protected areas has in fact barred local use of traditional forestlands. In others, land is not well demarcated and is subject to multiple ownership claims. Central government authorities and concession owners are also frequently

Land use conflicts

In one month during 2007, more than 500 conflicts were reported between local communities in Indonesia and private interests seeking to establish oil palm plantations (Friends of the Earth 2008). Commission of Pastors of the Land (CPT) in Brazil reported that 1 317 families were expelled from their land in 2006, with more than twice the number of families experiencing the same fate in 2007. CPT also reported that 19 people were assassinated over land rights issues in Brazil during 2007.

Local self-management comes with its own challenges, whether in the realm of protecting natural resources or establishing sustainable livelihoods. Key ingredients include reforming national laws and policies, investing in local forest governance, ensuring stakeholder involvement and raising awareness at the local and national levels.

➡ See also pages 14, 16, 32, 44

Certification
for sustainable forest management

Areas of forest certified as being under sustainable management have increased tremendously over the last ten years, but still only cover 7.6 per cent of the world's forests

In 1993, a number of environmental groups and other interested parties responded to growing concern about forest degradation and loss by creating the Forest Stewardship Council (FSC), a private initiative designed to promote voluntary forest certification. Over the following years the forest industry, forest owners associations and others created a number of additional certification organizations, including the Programme for the Endorsement of Forestry Certification (PEFC), the Canadian Standards Association, the Sustainable Forestry Initiative, Certificacao Florestal in Brazil, the Malaysia Timber Certification Council, Lembaga Ekolabel Indonesia, and the Chilean Forest Certification System.

Certification is a market-based mechanism designed to encourage environmentally sustainable and socially responsible forestry practices. Third party certification organizations use a range of environmental and social criteria to audit forestry operations and forest products. Timber operations awarded certificates can then use the certification label for marketing purposes.

The total area of forest falling under the major certification schemes has increased more than tenfold since 1998 (UNECE/FAO 2007). As awareness of the scale of global deforestation has grown, so has demand for certified forest products. Some large timber industry operators have used certification and best practices to establish a greener profile and expand their markets. The Canadian firm Tembec, for example, says certification helped the company to win the business of North America's biggest lumber buyer, Home Depot, and survive an industry downturn (Gazette 2008).

Certification has also benefited national economies. Bolivia's FSC-certified forests, soon to cover more than 2 million hectares (WWF-Bolivia 2008), make up about a quarter of the country's total forested area. As a result, Bolivia is one of the leading tropical forest nations to adopt the certification process. Officials say certification, backed by progressive forestry laws, has helped the country to export US$ 16 million worth of timber per year to the US and European markets. Certification has also provided important benefits to local communities, while encouraging a transition from the fell-

Source: FAO, FSC, PEFC. After a map compiled and produced by Moïse Tsayem Demaze, University of Maine, Le Mans, France.

⬇ Forest certification: regional breakdown

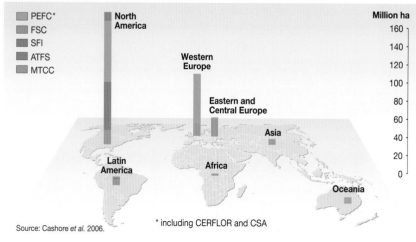

Source: Cashore *et al.* 2006.

* including CERFLOR and CSA

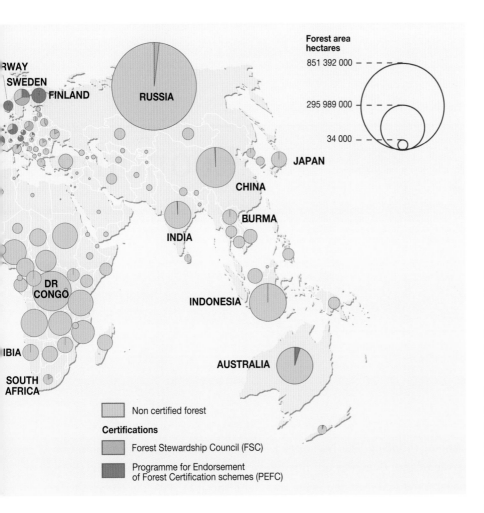

Forest area
hectares

851 392 000
295 989 000
34 000

NORWAY
SWEDEN
FINLAND
RUSSIA
JAPAN
CHINA
BURMA
INDIA
DR CONGO
INDONESIA
NAMIBIA
SOUTH AFRICA
AUSTRALIA

Non certified forest

Certifications

Forest Stewardship Council (FSC)

Programme for Endorsement
of Forest Certification schemes (PEFC)

← Very little forest area is certified

ing of mahogany to the use of more abundant species.

Despite such advances, however, only a small part of the world's forests – approximately 300 million hectares or 7.6 per cent of the total – has been certified (Forest Certification Resource Center 2008). PEFC certification accounts for approximately two-thirds of the total, with more than 200 million hectares of forests under its scheme. For the most part, certification has been concentrated in the boreal forests of the more developed northern countries. Some 56 per cent of the world's certified forest is located in North America, while approximately 34 per cent is in Europe and CIS (ITTO 2008).

The biggest challenge for certification is in developing countries, where deforestation of valuable tropical forests is occurring. Developing countries currently account for less than 10 per cent of the total area of certified forest globally, and half of that area is forest

plantation (Gulbrandsen 2006). Lack of awareness of certification programmes and a shortage of local technical capacity contribute to this geographic imbalance. Cost has also been identified as a barrier to certification, particularly for small landowners (i.e. 5-20 hectares) (Hansen *et al.* 2006) and community-based forest enterprises (Humphries and Kainer 2006) that are also unsure whether there will be a financial return on their certification investment. Some certifying organizations are attempting to address cost and access issues by enabling group certifications.

Meanwhile certification has come in for considerable criticism. Some schemes have been accused of allowing the fraudulent misuse of their eco-labels, of inadequate monitoring of profit-driven third-party certifying organizations, of accepting wood from mixed sources (i.e. certified and non-certified) and of certifying operations using unsustainable practices. More generally, certification has also been

accused of helping to fuel increased demand for forest products, instead of promoting the use of recycled wood or other alternative materials (FSC-Watch 2008).

In countries where government institutions and law enforcement mechanisms are weak, certification cannot be expected to work properly and to lead to sustainable management of forests. Indeed, degradation of the world's forests continues at an alarming rate. Some countries, such as Norway, have announced they simply will not procure any tropical wood to be used for public buildings, certified or not.

Nevertheless, increasing interest in green construction and procurement, and the use of biomass from forests for energy, suggests the demand for certified wood will continue to increase. The question is whether wood grown using truly sustainable and socially responsible practices will be able to meet this increased demand.

→ See also pages 24, 26, 58

↓ Trends in forest certification

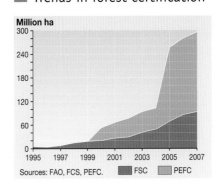

Million ha

Sources: FAO, FCS, PEFC.   ■ FSC   ▨ PEFC

⬇ Logging and corruption

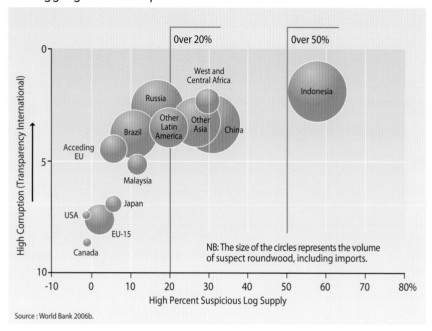

Source : World Bank 2006b.

Economic

In recent years, as awareness of forest ecosystem services has grown, economic incentives have been increasingly used to promote forest conservation

Despite the acknowledged importance of forests, government incentives have traditionally supported logging and the conversion of forests for agricultural purposes. Most forest nations have encouraged domestic forest operations by providing a combination of incentives including free or under-valued land rents, grants for harvesting, infrastructure and transport, interest-free loans, tax breaks and agricultural resettlement programmes (WRI 1988; OECD 2002). According to one estimate, some US$2 billion were granted each year in subsidies to industrial forest plantations. This is four times greater than the annual development assistance given to forest conservation (White 2006).

Yet government backed incentives aimed at conserving forests are increasing. In 1997, Costa Rica, recognizing the benefits that forest ecosystem services provide to society as a whole, began paying landowners to conserve or increase forest areas. The programme has helped restore much of Costa Rica's forest cover and is aiding the fight against poverty. A number of international environmental organ-

izations, including The Nature Conservancy and Conservation International, have used trust and estate laws to encourage gifts and bequests of land for conservation purposes. Conservation income tax credits and other tax incentives are also becoming increasingly popular in the United States and Europe (Shine 2005).

Through its loan policies and programmes, the World Bank has also played a significant role in forest conservation. In 2005 the World Bank announced it would extend its partnership with the World Wildlife Fund

(WWF) in the Alliance for Forest Conservation & Sustainable Use (Forest Alliance). In 2005, the Forest Alliance agreed to devote resources to reducing deforestation by 10 per cent by 2010 (WWF 2005). At the same time the World Bank has also been strongly criticized for funding forest projects that contribute to deforestation (Rainforest Foundation 2005; The Ecologist 2007).

Conserving forests has become a key weapon in the fight to reduce carbon emissions and slow climate change. According to the Intergovernmental Panel on Climate Change

⬇ When forest conversion is profitable...

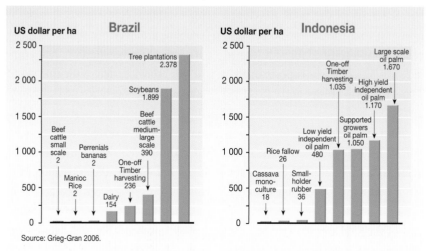

Source: Grieg-Gran 2006.

incentives to protect forests

(IPCC), deforestation is the cause of approximately 17 per cent of all greenhouse gases (GHG). At current rates of destruction, GHG emissions from deforestation in Brazil and Indonesia alone would equal approximately 80 per cent of the emissions reductions achieved under the Kyoto Protocol by 2012 (Santilli *et al.* 2005).

The concept of Reducing Emissions from Deforestation and Degradation (REDD), has gained strong support among environmental organizations and governments. In 2007, during the Conference of the Parties of the UN Framework Convention on Climate Change (UNFCCC), governments agreed to consider "policy approaches and positive incentives on issues relating to reducing emissions from deforestation and forest degradation (REDD) in developing countries; and the role of conservation, sustainable management of forests and enhancement of forest carbon stocks in devel-

⬇ Economic value of forests in the Mediterranean Basin

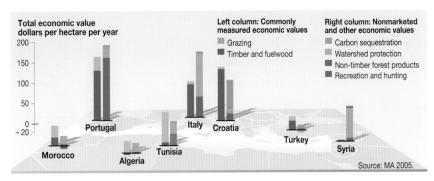

Source: MA 2005.

oping countries". Some nations such as Brazil pressed for direct payments to forest nations for protection of tropical forests. Others believed that units of forest should be assigned value, based on their ability to store carbon. These values should then be traded among willing buyers in the carbon credit market.

A number of REDD initiatives were announced during the UNFCCC conference. The World Bank launched its US$250 million Forest Carbon Partnership Facility, aimed at building capacity for REDD in developing countries and pioneering on a relatively small scale, performance-based incentive payments in pilot countries. The government of Norway also announced its intention to allocate US$2.7 billion over five years to prevent deforestation and reduce CO$_2$ emissions in developing countries. Norway subsequently announced its first partnership under this initiative with Tanzania.

Creating an effective REDD implementation mechanism will be politically and technically complex. One question still not answered is whether carbon

market trading is the best way to avoid further deforestation. Concerns have also been raised over whether the REDD regime can be implemented where there is inadequate governance, and there is concern that efforts to conserve forests might ignore community rights, in some cases resulting in displacement of forest inhabitants (FERN 2008).

In addition, REDD faces a number of significant technical challenges, including the establishment of effective programmes, accurately assessing forest carbon emissions and setting equitable reference emission levels. Concerns also exist regarding costs for measuring and monitoring deforestation and forest degradation, establishing what is and what is not permanent forest and on the issue of leakage – the possibility that forest protection zones will displace rather than eliminate deforestation.

As the search continues for ways to address climate change and protect forest ecosystems, the debate over these programmes and other innovative financial incentives will undoubtedly intensify.

⬇ Pricing ecosystems

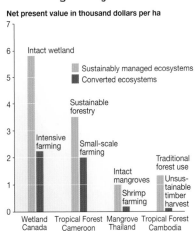

Source: MA 2005.

Are legal instruments sufficient

The fate of the world's forests, and of the rights of those who live in and around them, depends on effective governance, which includes negotiating among public and private stakeholders, and establishing and enforcing binding forest laws and policies.
(Rametsteiner 2007)

The quality of governance in regard to forests varies widely around the world. When governance is weak, forests are frequently subject to unplanned development and illegal logging, which in turn encourages illegal encroachment on forest land, illegal trade, crime, corruption and conflict.

The World Bank has estimated that the market value of illegal timber from public land is US$10 billion – more than eight times the total official development assistance given for sustainable forest management. In addition, governments lose an estimated US$5 billion more in unpaid tax revenue (World Bank 2006a).

The issue of illegal logging is increasingly recognized as an urgent problem by the world's governments. The G8 Action Plan on Forests includes actions to combat illegal logging. Several international organizations including the United Nations Food and Agriculture Organization (FAO), The United Nations Forum on Forests (UNFF), the Internatonal Tropical Timber Organization (ITTO) and the World Bank have been actively involved in programmes to combat illegal logging and associated trade.

⬇ **Global protected forests**

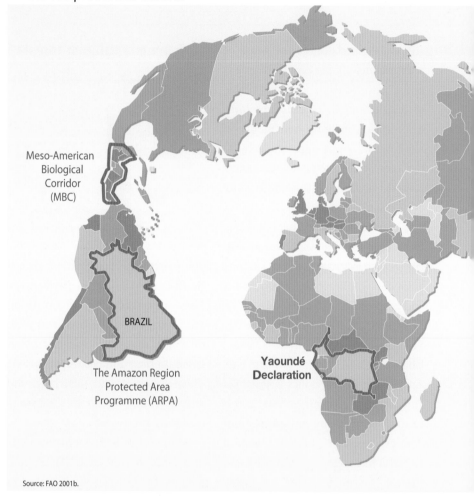

Meso-American Biological Corridor (MBC)

BRAZIL

The Amazon Region Protected Area Programme (ARPA)

Yaoundé Declaration

Source: FAO 2001b.

At the 1992 Rio Earth Summit, the issue of forests and deforestation was high on the political agenda. However, a convention on forests or a legally binding agreement similar to those on climate change, biological diversity and desertification, was not agreed (UNFF 2004). Intense negotiations at the Summit resulted in the Forest Principles, a set of non-legally binding targets for management, conservation and the sustainable development of forests. The Principles did not directly address such key issues as illegal logging and how to balance forest use with conservation (Gulbrandsen and Humphries 2006).

In 2007, the UN General Assembly agreed to a non-legally binding agreement that sets global standards for sustainable forest management and promotes integration of forest policies with other policies at national level (UN General Assembly 2007). It outlined future priorities in the form of the four Global Objectives on Forests. These call for reversal of the loss of forest cover; prevention of forest degradation; enhancing economic, social and environmental benefits from forests; increasing the area of protected forests and mobilizing finance for sustainable forest management. Countries have

to protect our forests?

« Heart of Borneo »
declaration

INDONESIA

Protected forest area in percentage of forest total area

- More than 20%
- 10 to 20%
- 5 to 10%
- Less than 5%
- Not protected
- Major legal large scale conservation mechanisms

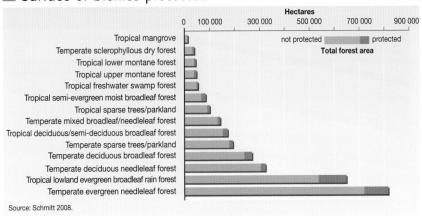

⬇ Surface of biomes protected

Source: Schmitt 2008.

committed to work towards these Global Objectives by 2015 (UN 2007).

In relation to protected areas, international organizations such as the World Wildlife (WWF), the International Union for Conservation of Nature (IUCN) and its World Commission on Protected Areas (WCPA) have done much to establish a representative network of protected areas around the world. As of 2006, countries had designated over 63 000 protected areas encompassing nearly 1 600 million hectares (UNEP-WCMC 2008), an area nearly the size of Russia. Some 43 500 protected areas of which the year

of establishment has not been reported are not reflected in these figures

At a regional level, governments and other stakeholders have begun to meet in a series of Forest Law Enforcement and Governance (FLEG) gatherings, covering East Asia and Pacific (2001), Africa (2003), and Europe and North Asia (2005). Similar initiatives have started in the Amazon and Central America. As of 2006, 90 producer and consumer countries were involved in FLEG processes.

In 2003 the European Union, a major importer of timber from the Amazon, Congo and northwest Russia regions, adopted an Action Plan for Forest Law Enforcement Governance and Trade (FLEGT), giving renewed emphasis to regulating the timber trade. FLEGT established a voluntary scheme to ensure that only legally harvested timber is imported into the EU from countries participating in the scheme (EU 2008). Though entry of timber into the EU is controlled via import licenses and bilateral FLEGT Voluntary Partnership Agreements (VPAs), concerns remain that VPA states could still import illegal timber into the EU via non-VPA countries (EU 2008). A number of EU

states (e.g., Belgium, Denmark, France, Germany and the UK) have created national procurement policies that require imported timber to be legally and sustainably harvested. Norway, a non-EU nation, has gone so far as to ban state procurement of any tropical timber.

International finance institutions and export credit agencies are also key multilateral players in ensuring forest governance. By lending to governments and providing government-guaranteed loans to corporations, these institutions are stakeholders in the forest trade sector and have a responsibility to perform due diligence to ensure that legitimate finance is not used for illegal activities.

As various governance processes have evolved, the governance landscape has grown ever more complex, with increasing numbers of transnational, civil society and other stakeholders influencing decisions. Strengthening local rule of law, land rights, transparency and participation opens up the potential for cooperation among all stakeholders and, with it, the possibility that sustainable forest management can be achieved.

Greening degraded forest landscapes

While there is no substitute for the environmental services provided by natural forests, as forests worldwide continue to disappear at an alarming rate, increased efforts are being made to reestablish forest areas

Measures aimed to create and reestablish forests can take a number of forms, including restoration, rehabilitation, reforestation and afforestation. In some countries such efforts have led to a change from net forest loss to net forest gain.

Successful forest landscape restoration starts from the ground up, with the close participation of local communities who often possess traditional forest-related knowledge. In Tanzania, the Shinyanga region used to have extensive acacia and

miombo woodland, until population pressures combined with agropastoral practices resulted in widespread deforestation and land degradation. Yet since 1986 the Shinyanga Soil Conservation Programme has promoted conservation of woods and grassland using the traditional indigenous natural resource management system *Ngitili* or fodder reserves. As a result, by 2000 between 378 000 and 472 000 hectares of woodlands had been restored in 833 villages in the Shinyanga region (Barrow *et al.* 2003).

⬇ Global forest plantations for protective purposes

Percentage of total forest area

- Over 25%
- 10 to 25%
- 1 to 10%
- Less than 1%
- No data available

Source: FAO 2006a.

⬇ Which countries account for the largest area of protective forest plantations?

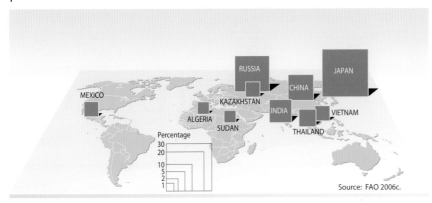

Source: FAO 2006c.

In 1999, China launched the Grain for Green programme to promote recovery of vegetation cover, watershed management and poverty alleviation through a grain and cash subsidy scheme. The programme now covers 25 provinces in over 1 600 counties, (autonomous regions and municipalities) and involves 15 million households and 60 million farmers. From 1999 to 2002, 7.7 million hectares of land was converted into forest, including 3.72 million hectares of farmland-turned forest and 3.98 million hectares of plantations established on barren hills. In 2002 alone, over 5 billion kg of grain and US$67 million in cash were disbursed to over 10 million farm households. The Grain for Green programme is considered the largest participatory community forest project in China, possibly in the world (Zhiyong 2003).

More recently, the planting of trees has also emerged as a vital tool for addressing climate change. Under the Kyoto Protocol, companies and governments can help meet their greenhouse gas emission caps by financing offset projects, such as tree planting designed to absorb CO_2. A voluntary carbon market is emerging in which people, companies and governments can purchase carbon offsets to compensate for greenhouse gas emissions caused, for example, by their use of transportation or electricity generation. Although the carbon market is growing fast, soaring in value from US$10.8 billion to US$64 billion between 2005-2007 (World Bank 2007), tree planting still represents only a relatively small percentage of the market (UNEP Riso Centre 2008). In addition, a number of concerns exist about the science and sustainability of offsetting carbon emissions through reforestation and afforestation (e.g., Carbon Trade Watch 2007; Grist 2007).

Forests also play an important role in adaptation to the effects of climate change. In 2008, India announced it would spend US$2.4 billion to restore six million hectares of degraded forests, not only to help absorb carbon emissions but also to stabilize shorelines and protect coastal human infrastructure in the face of rising sea levels (Sharma 2008).

In the aftermath of the 2004 Indian Ocean tsunami, one of the most devastating natural disasters in recent history, the role of mangroves and other coastal forests in the mitigation of tsunami impacts became a focus of international attention. Coastal forests of all types, including mangroves, beach forests and plantations, were found to absorb part of the tsunami's energy, hence reducing its impacts. At present mangrove-planting projects are gaining momentum as a means of establishing or restoring natural storm barriers and reducing disaster risks (UN ISDR 2008).

Agroforestry and the planting of trees with crops can play an important role in biodiversity conservation and food security. To date, the UN World Food Programme has supported the planting of 60 million trees in 35 countries, a development likely to play an important part in helping to fight the global food crisis (WAC 2008).

Efforts aimed at reforesting areas involve both big and small scale projects. An international group of musicians, music teachers and artisans have banded together to conserve pernambuco, a wood essential for the production of high quality bows for stringed instruments. The bow makers have collaborated with cacao farmers, scientists, landless people and government authorities to plant over 130 000 pernambuco seedlings on Brazil's heavily deforested Atlantic Coast (IPCI 2008).

Global partneship

In 2003, the Global Partnership on Forest Landscape Restoration was launched to act as a catalyst for a network of diverse examples of forest landscape restoration that benefit local communities and nature.

The partnership is a network of governments, organizations, communities and companies who together recognize the importance of forest landscape restoration and want to be part of a coordinated global effort. The Billion Tree Campaign launched by UNEP and the World Agroforestry Centre in 2006 planted more than 2 billion trees in 18 months, and has reset its goal to 7 billion trees planted. Further efforts such as these will be needed if the tide of global forest destruction is to be reversed.

→ See also pages 6, 10, 32, 56

Glossary	
Restoration	Recreating most of the ecological functions of the original forest on land that was forested in the past.
Rehabilitation	Recreating some ecological functions on land that was forested in the past.
Reforestation	Planting (or sowing) of trees on land which was forested in the recent past.
Afforestation	Planting (or sowing) of trees on land that has not been forested before (or at least not within the last 50 years).

References

ABC, UNEP, NOAA (2002). Project Atmospheric Brown Cloud. The integration of air pollution and climate science to assess impacts on environment and society. Available at http://www-abc-asia.ucsd.edu/ abcbromastercopyimprints.pdf

Achard F., Eva H.D., Mollicone D., Beuchle R. (2008). The effect of climate anomalies and human ignition factor on wildfires in Russian boreal forests. Philosophical Transactions of the Royal Society B 363: 2331–2339.

AFP (2008). Brazil wants 21 billion dollars to save the Amazon. 1 August 2008. Agence France-Presse.

AMAP (1998). AMAP assessment report: arctic pollution issues. Arctic Monitoring and Assessment Programme, Oslo.

Amazon Institute for Environmental Research, Woods Hole Research Center, Federal University of Minas Gerais (2006). The Amazon in a changing climate: large-scale reductions of carbon emissions from deforestation and forest impoverishment. Available at http:// www.whrc.org/pressroom/press_ releases/pr-2006-11-17-amazon-climate-chg.htm

Angelsen A., Wunder S. (2003). Exploring the forest-poverty link: key concepts, issues and research implications. Center for International Forestry Research Occasional Paper No. 40. Bogor.

Barrow E., Kaale B., Mlenge W. (2003). Forest landscape restoration in Shinyanga, the United Republic of Tanzania. Available at http://www. fao.org/docrep/ article/wfc/xii/0164-b3.htm

BC Ministry of Environment (2000). Just the facts – A review of silviculture and other forestry statistics. British Columbia Ministry of Environment, Province of British Columbia.

Beniston M. (Ed.) (1994). Mountain environments in changing climates. Routledge Publishing Co., London and New York.

Billington C., Kapos V., Edwards M.S., Blyth S., Iremonger S. (1996). Estimated original forest cover map – a first attempt. WCMC, Cambridge, UK.

Braatz S., Fortuna S., Broadhead J., Leslie R. (Eds.) (2007). In: RAP Publication (FAO), No. 2007/07; Regional technical workshop on coastal protection in the aftermath of the Indian Ocean tsunami: What role for forests and trees?, Khao Lak 28–31 August 2006, FAO Regional Office for Asia and the Pacific, Bangkok.

Britton B., Stephens K.R., Gurney P., Tans P., Sweeney C., Peters W., Bruhwiler L., Ciais P., Ramonet M., Bousquet P., Nakazawa T., Aoki S., Machida T., Inoue G., Vinnichenko N., Lloyd J., Jordan A., Heimann M., Shibistova O., Langenfelds R.L., Steele L.P., Francey R.J., Denning A.S. (2007) Weak northern and strong tropical land carbon uptake from vertical profiles of atmospheric CO_2. Science 316(5832): 732.

Broadhead J., Bahdon J., Whiteman, A. (2003). Past trends and future prospects for the utilization of wood for energy: Annexes 1 and 2. Global Forest Products Outlook Study Working Paper No. GFPOS/WP/05. FAO, Rome.

Brown S. (2008). Beetle tree kill releases more carbon than fires. 23 April 2008. Nature News Available at http://www.nature.com/news/2008

Bubb P., May I., Miles L., Sayer J. (2004). Cloud forest agenda. UNEP-WCMC, Cambridge.

Carbon Trade Watch (2007). Carbon neutral myth – offset indulgences for your climate sins. Transnational Institute. Available at http://www.tni. org/detail_pub.phtml?&know_id=56

Case J.W. (1980). The influence of three sour gas processing plants on the ecological distribution of epiphytic lichens in the vicinity of Fox Creek and Whitecourt, Alberta, Canada. Water, Air and Soil Pollution 14(1), 45–68.

Cavendish W. (2000). Empirical regularities in the poverty-environment relationship of rural households: evidence from Zimbabwe. World Development 28(11): 1979–2003.

Cashore B. Gale F., Meidinger E., Newsom D. (Eds.) (2006). Confronting sustainability: forest certification and developing and transitioning countries. Yale School of Forestry and Environmental Studies Publication Series, Yale University, New Haven.

& internet links

CBFP (2006). The forests of the Congo Basin: state of the forest 2006. Congo Basin Forest Partnership. Available at http://www.cbfp.org

Chagnon F.J.F. and Bras R.L. (2005). Contemporary climate change in the Amazon. Geophysical Research Letters 32: L13703.

Colchester M., Jiwan N., Andiko, Sirait M., Firdaus A.Y., Surambo A., Pane, H. (2006). Promised land. Palm oil and land acquisition in Indonesia – implications for local communities and indigenous peoples. Forest Peoples Programme, Perkumpulan Sawit Watch, HuMA and the World Agroforestry Centre, UK.

Collier P. (2007). The bottom billion. Oxford University Press, Oxford.

Comissão Pastoral da Terra (2008). Web page at http://www.cptnac.com.br/?system=news&eid=6

CBD (2004). 7th Conference of the Parties to the Convention on Biological Diversity. Kuala Lumpur 9–20 February 2004. Available at http://www.cbd.int/doc/handbook

Cossalter C., Pye-Smith C. (2005). Fast-wood forestry: myths and realities. Forest Perspectives No. 1. Center for International Forestry Research, Bogor.

Crutzen P. J., Mosier A.R., Smith K.A., Winiwarter W. (2008). N2O release from agro-biofuel production negates global warming reduction by replacing fossil fuels. Atmospheric Chemistry and Physics 8: 389–395.

Cunningham A.B. (2001). Applied ethnobotany. Earthscan, London.

Curran L.M., Trigg S.N., McDonald A.K., Astiani D., Hardiono Y.M., Siregar P., Caniago I., Kasischke E. (2004). Lowland forest loss in protected areas of Indonesian Borneo. Science 303(5660).

Dadzie R. (2006–2007). Neoclassical versus institutionalist views on land tenure systems: implications for economic development. Oeconomicus, Vol VIII, 2006–2007.

Debroux L., Hart T., Kaimowitz D., Karsenty A., Topa G. (2007). Forests in post-conflict Democratic Republic of Congo: analysis of a priority agenda. CIFOR/World Bank, Washington DC.

Dobson A., Lynes L. (2008). How does poaching affect the size of national parks? Trends in Ecology & Evolution, Vol 23(4): 177–180.

Down to Earth (2006). No chip mill without wood: a study of UFS projects to develop wood chip and paper pulp mills in Kalimantan, Indonesia. Available at http://dte.gn.apc.org/cskal06.pdf

Earth Trends Database (2008). World Resources Institute. Available at http://earthtrends.wri.org/

Ecologist Online (2007). World Bank: deforestation good for growth. Available at http://www.theecologist.org/pages/archive_detail.asp?content_id=760

EEA (2006). Progress towards halting the loss of biodiversity by 2010. European Environment Agency, Copenhagen.

Eickhout B., van den Born G.J., Notenboom J., van Oorschot M., Ros J.P.M., van Vuuren D.P., Westhoek H.J. (2008). Local and global consequences of the EU renewable directive for biofuels. Testing the sustainability criteria. MNP, Bilthoven.

Ellis F. (2000). Rural livelihoods and diversity in developing countries. Oxford University Press, Oxford.

EOSL/CCT/FONAFIFO (2002). Estudio de cobertura forestal de Costa Rica con imágenes Landsat Tm 7 para el año 2001. Earth Observation Systems Laboratory (EOSL), University of Alberta, Canada; Centro Científico Tropical and Fondo Nacional de Financiamento Forestal, Costa Rica. Available at http://documentacion.sirefor.go.cr/archivo/cobertura/INFORME_COBERTURA_97_00.pdf

EU (2008). Forest law enforcement, governance and trade (FLEGT). Website: http://ec.europa.eu/environment/forests/flegt.htm

FAO (2001a). State of the World's Forests 2001. FAO, Rome.

FAO (2001b). Global Forest Resources Assessment 2000 – Main report. FAO Forestry Paper 140. FAO, Rome.

FAO (2005). FAO and the challenge of the millennium development goals: The road ahead. Discussion paper. FAO, Rome.

FAO (2006a). Global Forest Resources Assessment 2005 – Progress towards sustainable forest management. FAO Forestry Paper 147. FAO, Rome.

FAO (2006b). Global planted forest thematic study, results and analyses. By Del Lungo A., Ball J., Carle J. Planted Forests and Trees Working Paper No. 38. FAO, Rome.

FAO (2006c). Livestock's long shadow: Environmental issues and options. FAO, Rome.

FAO (2007). State of the World's Forests 2007. FAO, Rome.

FAO (2008a). FAOSTAT database. Available at http://faostat.fao.org

FAO (2008b). Forests and energy – key issues. FAO Forestry Paper 154. FAO, Rome.

FAO (2008c). Forests and water, by L. Hamilton. FAO Forestry Paper 155. FAO, Rome.

FAO (2009). State of the World's Forests 2009. FAO, Rome (In press).

Fargione J., Hill J., Tilman D., Polasky S., Hawthorne P. (2008). Land clearing and the biofuel carbon debt. Science 319: (5867).

FERN (2008). Response to the Eliasch Review questionnaire: global forests and finance flows. Available at http://www.fern.org/media/documents/document_4131_4132.pdf

FO Licht World Ethanol and Biofuels Report (2008). Available at http://www.agra-net.com

FONAFIFO (2007). Estudio de monitoreo de cobertura forestal de Costa Rica 2005. Fondo Nacional de Financiamiento Forestal; EOSL, Universidad de Alberta. Available at http://www.sirefor.go.cr/coberturaforestal.html

Forbes K., Broadhead J. (2008). The role of coastal forests in the mitigation of tsunami impacts. RAP publication 2007/1. FAO Regional Office for Asia and the Pacific, Bangkok.

Forest Certification Resource Center (2008). Website at http://www.metafore.org/index.php?p=Forest_Certification_Resource_Center&s=147

FORUM (2007). Beyond formalisation: a land fights agenda for Norwegian development and foreign policy. The Norwegian Forum for Environment and Development, Oslo.

Franky C., Mahecha D. (2000). La territorialidad entre los pueblos de tradición nómada del noroeste amazónico colombiano. In Territorialidad Indígena y Ordenamiento de la Amazonía. Universidad Nacional de Colombia,Fundación GAIA Amazonas, Santa Fé de Bogotá.

Friends of the Earth (2008). Losing ground. The human rights impacts of oil palm plantation expansion in Indonesia. Friends of the Earth, LifeMosaic and Sawit Watch.

FSC (2008). Plantations Review. Available at http://www.fsc.org/plantationsreview.html

FSC-Watch (2008). Website at http://www.fsc-watch.org

Gazette (2008). Going green helped us survive. 28 February 2008. The Gazette.

Grainger A. (1993). Controlling tropical deforestation. Earthscan, London.

Greenpeace (2007). Carving up the Congo. Available at http://www.greenpeace.org/international/campaigns/forests/africa/congo-report

Grieg-Gran M. (2006). The cost of avoiding deforestation. Report prepared for the Stern Review of the economics of climate change. International Institute for Environment and Development, London.

Griffiths T. (2007). Seeing 'REDD'? 'Avoided deforestation' and the rights of indigenous peoples and local communities. Forest Peoples Programme, UK.

Grist (2007). Carbon offsets: The real reasons to avoid trees. 10 July 2007. Grist. Available at http://www.grist.com

Gulbrandsen L., Humphreys D. (2006). International initiatives to address tropical timber logging and trade. The Fridtjof Nansen Institute, Norway.

Hanes S. (2006). Loss of trees, loss of livelihood. Pulitzer Center on Crisis Reporting. Available at http://www.pulitzercenter.org/loss.htm.

Hansen E., Fletcher R., Cashore B., McDermott C. (2006). Forest certification in North America. Oregon State University Extension Service. Available at http://www.yale.edu/forestcertification/pdfs/2006/OSU_SFI-CertComparStudy.pdf

Hansen M.C., DeFries R.S., Townshend J.R.G., Carroll M., Dimiceli C., Sohlberg R.A. (2003). Global percent tree cover at a spatial resolution of 500 meters: first results of the MODIS vegetation continuous fields algorithm. Earth Interactions, Volume 7(10): 1–15.Houghton R.A. (2003). Revised estimates of the annual net flux of carbon to the atmosphere from changes in land use and land management 1850-2000. Tellus 55B:378–390.

Hooijer A., Silvius M., Wösten H., Page S. (2006). PEAT-CO2, Assessment of CO2 emissions from drained peatlands in SE Asia. Delft Hydraulics Report Q3943.

Humphries S., Kainer K. (2006) Local perceptions of forest certification for community-based enterprises. Forest Ecology and Management, 235(1–3): 30–43.

IEA (2002). Energy and Poverty. Chap[ter 13 in: World Energy Outlook 2002. International Energy Agency, Paris.

IEA (2004). Biofuels for transport: an international perspective. Available at http://www.iea.org/textbase/nppdf/free/2004/biofuels2004.pdf

INPE (2008). National Institute for Space Research. Website at http://www.obt.inpe.br/prodes

International Pernambuco Conservation Initiative. (2008) Website at http://www.ipci-usa.org

ITTO (2008). Developing forest certification. Towards increasing the comparability and acceptance of forest certification systems worldwide. ITTO Technical Series 29, International Tropical Timber Organization, Yokohama.

Jouzel J., Debroise A. (2007). Le climat: jeu dangereux : Dernières nouvelles de la planète, Editions Dunod, Paris.

Kabat P., Claussen M., Dirmeyer P.A., Gash J.H.C., de Guenni L.B., Meybeck M., Vorosmarty C.J., Hutjes R.W.A., Lütkemeier S. (Eds.) (2004). Vegetation, water, humans and the climate: A new perspective on an interactive system. The International Geosphere-Biosphere Programme Series, Springer Verlag, Heidelberg.

Kanninen M., Murdiyarso D., Seymour F., Angelsen A., Wunder S., German L. (2007). Do trees grow on money? The implications of deforestation research for policies to promote REDD. Center for International Forestry Research, Bogor.

Kersanty A. (2006). Les enjeux méconnus de l'économie du bois. July 2006. Le Monde Diplomatique.

Kinver M. (2008). Mangrove loss 'put Burma at risk'. 6 May 2008. BBC News.

Kirkup P. (2001). Global biodiversity scenarios for the year 2050: application of species-area relationships to assess the impact of deforestation on the diversity of tree species. Master of Research thesis, University of Edinburgh.

Kozlov M.A., Barcan V. (2000). Environmental contamination in the central part of the Kola Peninsula: history, documentation and perception. Ambio 29(8), 512–517.

Krogh A.C. (2006). Med jaguarens kraft. Bazar Forlag AS, Oslo.

Kusters K., Achdiawan R., Belcher B. and Pérez M.R. (2006). Balancing development and conservation? An assessment of livelihood and environmental outcomes of non-timber forest product trade in Asia, Africa and Latin America. Ecology and Society 11(2): 20.

Larson A.M., Ribot J.C. (2007). The poverty of forestry policy: double standards on an uneven playing field. Sustainability Science 2(2): 189–204.

Lund H.G. (2008). Definitions of forest, deforestation, afforestation and reforestation. Forest Information Services. Available at http://home.comcast.net/~gyde/DEFpaper.htm

MA (2005). Ecosystems and human well-being: current state and trends. Millennium Ecosystem Assessment Series, Vol 1. Island Press, Washington DC.

Malhi Y., Timmons Roberts J., Betts R.A., Killeen T.J., Li W., Nobre C.A. (2008). Climate change, deforestation, and the fate of the Amazon. Science 319(5860): 169.

McGregor J. (1995). Gathered produce in Zimbabwe's communal areas: changing resource availability and use. Ecology of Food and Nutrition 33(3): 163–193.

MCPFE (2005). Combating illegal harvesting and related trade of forest products in Europe. Report for the MCPFE Workshop held in Madrid, Spain 3–4 November 2005.

MCPFE (2007). State of Europe's Forests 2007. Jointly prepared by the MCPFE Liaison Unit, UNECE/FAO. Warsaw.

Medlyn B., McMurtrie R., Dewar R., Jeffreys M. (2000). Soil processes dominate the long-term response of forest net primary productivity to increased temperature and atmospheric CO_2 concentration. Canadian Journal of Forest Research, 30: 873–888.

Miles L., Grainger A., Phillips O. (2004). The impact of global climate change on tropical forest biodiversity in Amazonia. Global Ecology and Biogeography 13: 553–565.

MNRO (2008). Ontario's forests – harvesting in the boreal forests. Ministry of Natural Resources, Ontario, Canada.

Mollicone D., Eva H.D., Achard F. (2006). Ecology: human role on Russian wild fires. Nature (Brief communication) 440: 436–437.

Molnar A. (2003). Forest certification and communities: looking forward to the next decade. Forest Trends, Washington DC.

Mongabay.com. Website at: http://www.mongabay.com/brazil.html

Mountinho P., Schwartzman S. (Eds.) (2005). Tropical deforestation and climate change. Amazon Institute for Environmental Research, Belém, Pará, Brazil.

Moyini Y., Mosiga M., Byaruhanga A., Ssegawa P. (2008). Economic valuation of the proposed degazettment of part of Mabira Central Forest Reserve (In press).

Narain U., Gupta S., Veld K.V. (2008). Poverty and resource dependence in rural India. Ecological Economics 66: 161–176.

Nepstad D., Stickler C., Almeida O. (2006). Globalization of the Amazon beef and soy industries: opportunities for conservation. Conservation Biology 20(6): 1595–1603.

Neumann R.P., Hirsch E. (2000). Commercialisation of non-timber forest products: review and analysis of research. Center for International Forestry Research, Bogor, Indonesia.

Odegard J.T., Diss L., Løvold L. (2006). Indigenous peoples and collective land rights – urgent need for international recognition. In Legal empowerment: a way out of poverty. Brøther M.E. and Solberg J.A. (Eds.). Norwegian Ministry of Foreign Affairs, Oslo.

OECD (2002). Proceedings of the OECD Workshop on environmentally harmful subsidies. 7–8 November. Organization for Economic Co-operation and Development, Paris.

Oglethorpe J., Ericson J., Bilsborrow R.E., Edmond J. (2007). People on the move: reducing the impacts of human migration on biodiversity. World Wildlife Fund and Conservation International Foundation, Washington DC .

Olsen V. (2006). Protecting indigenous peoples' lands and resources: the role of the Constitutional Court of Colombia. Human Rights Everywhere. Available at http://www.hrev.org

Olson D.M., Dinerstein E., Wikramanayake E.D., Burgess N.D., Powell G.V.N., Underwood E.C., D'Amico J.A., Itoua I., Strand H.E., Morrison J.C., Loucks C.J., Allnutt T.F., Ricketts T.H., Kura Y., Lamoreux J.F., Wettengel W.W., Hedao P., Kassem K.R. (2001). Terrestrial ecoregions of the world: A new map of life on earth. BioScience 51: 933–938.

Pandit B.H., Albano A., Kumar C. (2008). Improving forest benefits for the poor: learning from community-based forest enterprises in Nepal. Center for International Forestry Research, Bogor.

Parish F., Sirin A., Charman D., Joosten H., Minayeva T., Silvius M., Stringer L. (2008). Assessment on peatlands, biodiversity and climate change. Global Environment Centre and Wetlands International, Kuala Lumpur-Wageningen.

Parmesan C., Yohe G. (2003). A globally coherent fingerprint of climate change impacts across natural systems. Nature 421(6918): 37–42.

Peksa-Blanchard M., Dolzan P., Grassi A., Heinimö J., Junginger M., Ranta T., Walter A. (2007). Global wood pellets markets and industry: policy drivers, market status and raw material potential. IEA Bioenergy Task 40. Available at http://bioenergytrade.org

Pitman N.C., Jørgensen P.M. (2002). Estimating the size of the world's threatened flora. Science, 298(5595): 989.

Povos Indígenas (2008). Fundação Nacional do Indio, Ministério da Justiça. Web site at: http://www.funai.gov.br

Rainforest Foundation UK (2005). Broken promises – how World Bank Group policies fail to protect forests and forest peoples' rights. Available at http://www.forestpeoples.org

REN21 (2006). Renewables global status report 2006 update. REN21 Secretariat and Worldwatch Institute, Washington DC.

Reuters (2008) Amazon deforestation seen surging. 16 January 2008. Reuters.

Righelato R. and Spracklen D.V. (2007). Carbon mitigation by biofuels or by saving and restoring forests? Science 317(5840): 902.

Ruckstuhl K.E, Johnson E.A., Miyanishi K. (2007). Introduction. The boreal forest and global change. Philosophical Transactions of the Royal Society B 363(1501): 2245–2249.

Sánchez E., Pardo M., Flores M., and Ferreira P. (2000). Protección del conocimiento tradicional elementos conceptuales para una propuesta de reglamentación: El caso de Colombia. Instituto de Investigación de Recursos Biológicos, Alexander von Humboldt, Santafé de Bogotá.

Santilli M., Moutinho P., Schwartzman S., Nepstad D., Curran L., Nobre C. (2005). Tropical deforestation and the Kyoto Protocol: an editorial essay. Climatic Change 71: 267–276.

Scherr S., White A., Kaimowitz D. (2003). A new agenda for forest conservation and poverty reduction: Making forest markets work for low-income producers. Forest Trends, Washington DC.

Schmitt C.B., Belokurov A., Besançon C., Boisrobert L., Burgess N.D., Campbell A., Coad L., Fish L., Gliddon D., Humphries K., Kapos V., Loucks C., Lysenko I., Miles L., Mills C., Minnemeyer S., Pistorius T., Ravilious C., Steininger M., Winkel G. (2008). Global ecological forest classification and forest protected area gap analysis. Analyses and recommendations in view of the 10% target for forest protection under the Convention on Biological Diversity (CBD). Freiburg University Press, Freiburg.

Scoones I., Melnyk M., Pretty J.N. (1992). The hidden harvest: wild foods and agricultural systems. IIED, London.

Sharma S. (2008). India plans to spend $2.4 billion on rejuvenating forestland. 8 May 2008. Bloomberg News. Available at http://www.bloomberg.com

Shine C. (2005). Using tax incentives to conserve and enhance biodiversity in Europe. Nature and Environment No. 143. Council of Europe Publishing, Strasbourg.

Statistics Canada (2008). Website at http://www12.statcan.ca

Stern N. (2006). Stern review of the economics of climate change. Office of Climate Change, UK.

Stickler C., Coe M., Nepstad D., Fiske G., Lefebvre P. (2007). Reducing emissions from deforestation and forest degradation (REDD): Readiness for REDD – a preliminary global assessment of tropical forested land suitability for agriculture. A Report for the United Nations Framework Convention on Climate Change (UNFCCC) Conference of the Parties (COP), Thirteenth Session, 3–14 December 2007, Bali. Woods Hole Research Center, Falmouth, MA.

Taiga Rescue Network (2008). Website at http://www.taigarescue.org

Taylor S.W., Carroll A.L., Alfaro R.I., Safranyik L. (2006). Forest, climate and mountain pine beetle outbreak dynamics in western Canada; in The Mountain Pine Beetle: A synthesis of biology, management, and impacts on lodgepole pine. Safranyik L. and Willson B. (Eds.). Natural Resources Canada, Canadian Forest Service, Pacific Forestry Centre, 67–94.

UN (2003). UN Security Council Resolutions. The situation in Liberia. Available at http://www.un.org/docs/sc/unsc_resolutions03.html

UN (2006). UN Security Council Resolution 1689. Security Council extends measures to prevent import of rough diamonds from Liberia; chooses not to renew measure on import of timber products. Available at http:// www.un.org/news/press/docs/2006/ sc8756.doc.htm

UNCTAD (2008). United Nations Conference on Trade and Development website available at http://www.unctad.org

UNECE/FAO (2007). Forests products annual markets review 2006–2007, Geneva Timber and Forest Study Paper 22. UN Economic Commission for Europe/FAO, Geneva.

UNEP (2001). An assessment of the status of the world's remaining closed forests. Division of Early Warning and Assessment, United Nations Environment Programme, Nairobi.

UNEP (2002). The Great Apes – The Road Ahead. UNEP World Conservation Monitoring Centre, Cambridge.

UNEP (2003). Afghanistan: post-conflict environmental assessment. Post-Conflict and Disaster Management Branch, United Nations Environment Programme, Geneva.

UNEP (2005). World Atlas of Great Apes and Their Conservation. United Nations Environment Programme's World Conservation Monitoring Centre, Cambridge.

UNEP (2006). The last stand of the Orang-utan – state of emergency. Illegal logging, fire and palm oil in Indonesia's national parks. UNEP Rapid Response Assessment, UNEP GRID-Arendal, Arendal.

UNEP GLOBIO (2008). Website at http://www.globio.info

UNEP/GRID-Arendal (2008). Web page at http://www.grida.no/publications/vg/climate/page/3081.aspx

UNEP Riso Centre (2008) Website at http://cdmpipeline.org/cdm-projects-type.htm

UNEP-WCMC (2008). Website at http://www.unep-wcmc.org/protected_areas/protected_areas.htm

UNFF (2004a). Fact Sheet 1: The United Nations Forum on Forests. UN Forum on Forests, New York.

UNFF (2004b). Fact Sheet 6: International trade. UN Forum on Forests, New York.

UNFF (2004c). Fact Sheet 7: Illegal logging and associated trade. UN Forum on Forests, New York.

UN General Assembly (2007). Non-legally binding instruments for all forests. A/RES/62/98. http://www.un.org/esa/forests/ nlbi-GA.html

UN ISDR (2008). International Strategy for Risk Reduction. Website address: www.unisdr.org/eng/risk-reduction/climate-change/ cc-adaptation.html

UNODC (2005). World Drug Report 2005. United Nations Office on Drugs and Crime, Vienna.

UPNG Remote Sensing Centre (2008). Website address: http://gis.mortonblacketer.com. au/upngis

Urgewald (2007). Banks, pulp and people: a primer on upcoming international pulp projects. Urgewald e.V., Germany.

USDA (2008). USDA Agricultural Projections to 2017. Long-term projections report, United States Department of Agriculture, Washington DC.

Väisänen R. (1996). Boreal forest ecosystems. IUFRO-95 Papers and Abstracts. Presented at IUFRO XX World Congress, 6–12 August 1995, Tampere.

Vedeld P., Angelsen A., Sjaastad E., Berg G.K. (2004). Counting on the environment – forest incomes and the rural poor. Environmental Economics Series Paper 98, Environment Department, World Bank, Washington DC.

WAC (2008). Billion tree campaign to grow into the seven billion tree campaign. Press release, 13 May 2008. World Agroforestry Centre, Available at http://www.worldagroforestry.org

Watson R.T., Zinyowera M.C., Moss R.H. (Eds.) (1995). Climate change 1995: impacts, adaptations and mitigation of climate change: scientific-technical analyses. Contribution of Working Group II to the second assessment of the Intergovernmental Panel on Climate Change. Cambridge University Press, Cambridge.

White A., Bull G.Q., Maginnis S. (2006). Subsidies for industrial plantations: turning controversy into opportunity. Aborvitae 31: 15.

Wood D.B. (2007). California's age of megafires. 24 October 2007. The Christian Science Monitor.

Woods Hole Research Center (2007). Three essential strategies for reducing deforestation. A report for the United Nations Framework Convention on Climate Change (UNFCCC) Conference of the Parties (COP), Thirteenth session, 3–14 December 2007, Bali, Indonesia. Woods Hole Research Center, Falmouth, MA.

World Bank (2004). Sustaining forests: a development strategy. World Bank, Washington DC.

World Bank (2006a). At loggerheads? Agricultural expansion, poverty reduction, and environment in the tropical forests. World Bank, Washington DC.

World Bank (2006b). Environment matters at the World Bank: 2006 annual review. World Bank, Washington DC.

World Bank (2006c). Global issues for global citizens: an introduction to key development challenges, Bhargava V.K. (Ed.) World Bank Publications, Washington DC.

World Bank (2007). Forests and forestry. Website at http://www.worldbank.org/fleg

WRI (1988). The forest for the trees? Government policies and the misuse of forest resources. World Resources Institute, Washington DC

WRI (2008a). Ecosystem services: a guide for decision makers. World Resources Institute, Washington DC.

WRI (2008b). Trees in the greenhouse. Why climate change is transforming the forest products business. World Resources Institute, Washington DC.

WWF (2005). International initiatives to address tropical timber logging and trade at failing the forests – Europe's illegal timber trade. Available at http://assets.panda.org/downloads/ failingforests.pdf

WWF (2008). Arctic climate impact science – an update since ACIA. Worldwide Fund for Nature, Gland.

WWF-Bolivia (2008). Website at http://www.panda.org/about_wwf/where_we_work/latin_america_and_caribbean/publications

Young B.E., Stuart S.N., Chanson J.S., Cox N.A., Boucher T.M. (2004). Disappearing jewels: the status of New World amphibians. NatureServe, Arlington, Virginia.

Zhiyong L. (2003). A policy review on watershed protection and poverty alleviation by the Grain for Green Programme in China. In Proceedings of the workshop on forests for poverty reduction: opportunities with CDM, environmental services and biodiversity. RAP Publication 2004/22, FAO Regional Office for Asia and the Pacific, Bangkok.